Partial Differential Equations for Geometric Design

Hassan Ugail

Partial Differential Equations Equations for Geometric Design

 Springer

Hassan Ugail
University of Bradford
Bradford
UK
H.Ugail@Bradford.ac.uk

ISBN 978-1-4471-6112-7   ISBN 978-0-85729-784-6 (eBook)
DOI 10.1007/978-0-85729-784-6
Springer London Dordrecht Heidelberg New York

British Library Cataloguing in Publication Data
A catalogue record for this book is available from the British Library

Cover design: VTeX UAB, Lithuania

Printed on acid-free paper

Springer is part of Springer Science+Business Media (www.springer.com)

Preface

This book is based on the results of over 14 years of research into the topic of partial differential equations applied to problems relating to geometric design. The book is intended as an introduction to the topic. It will equally serve as a reference for the mathematical fundamentals and modern applications using partial differential equations as a tool for geometric design. The book starts off with a gentle introduction to the relevant mathematical concepts for geometric design and then introduces partial differential equations to the reader.

The bulk of the book relates to the use of a class of partial differential equations known as elliptic partial differential equations which are used for surface generation, manipulation as well as design for function. Throughout the book, in order to enhance the understanding of the reader, practical examples with relevant illustrations as well explanations are used extensively. Moreover, for the purpose of enabling the reader to gain practical experience some examples of computer code is supplied.

The author gratefully acknowledges his appreciation and gratitude to various colleagues who have collaborated with him on the research relevant to this book. He also gratefully acknowledges the research funding he has received over the years from various UK research funding agencies, especially from the UK Engineering and Physical Sciences Research Council (EPSRC). The author also acknowledges the help and support he has received by many of his research assistants and research students who have been involved in research relating to the topic of this book and consequently contributed to this book both directly and indirectly.

Ilkley, UK   Hassan Ugail

Contents

Chapter 1
Elementary Mathematics for Geometric Design

Abstract This chapter deals with some of the basic mathematical concepts that are required to fully understand the material discussed in the rest of this book. Particularly, this chapter presents, in a concise form, the basic concepts of vector algebra, matrices, systems of linear equations and mathematical properties of surfaces.

1.1 Vector Algebra

A point in space is usually defined by providing its location relative to three mutually perpendicular coordinate axes passing through an origin O. This is represented as x, y, z axes as shown in Fig. 1.1. A point P is said to have rectangular coordinates (x, y, z) if

- x is its signed distance from the yz plane,
- y is its signed distance from the xz plane,
- z is its signed distance from the xy plane.

We define the three dimensional space $\mathbf{R}^3$ to be the set of all triples (x, y, z) of real numbers. These elements of $\mathbf{R}^3$ are called vectors. Thus, the point $P(x, y, z)$ defines the vector $\mathbf{v} = (x, y, z)$ in $\mathbf{R}^3$. The vector $\mathbf{v}$ can also be denoted as

$$\mathbf{v} = \begin{bmatrix} x \\ y \\ x \end{bmatrix}.$$

Given two vectors $\mathbf{a} = (\mathbf{a_1}, \mathbf{a_2}, \mathbf{a_3})$ and $\mathbf{b} = (\mathbf{b_1}, \mathbf{b_2}, \mathbf{b_3})$, their sum vector is defined as

$$\mathbf{a} + \mathbf{b} = (a_1 + b_1, a_2 + b_2, a_3 + b_3).$$

The displacement vector $\mathbf{v}$ with a starting point (a_1, a_2, a_3) and final point (b_1, b_2, b_2) is defined as

$$\mathbf{v} = (a_1 - b_1, a_2 - b_2, a_3 - b_3).$$

If c is a real number, then the scalar product of $\mathbf{a}$ with c is the vector

$$c\mathbf{a} = (ca_1, ca_2, ca_3).$$

H. Ugail, *Partial Differential Equations for Geometric Design*,
DOI 10.1007/978-0-85729-784-6_1, © Springer-Verlag London Limited 2011

Fig. 1.1 Representation of
vector **v** in $\mathbf{R}^3$

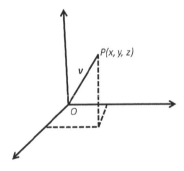

The length $|a|$ of **a** is defined to be

$$|a| = \sqrt{(a_1)^2 + (a_2)^2 + (a_3)^2}.$$

Consider any three vectors $\mathbf{u} = (u_1, u_2, v_2)$, $\mathbf{v} = (v_1, v_2, v_3)$ and $\mathbf{w} = (w_1, w_2, w_3)$ in $\mathbf{R}^3$. **u**, **v** and **w** are said to be linearly dependent if and only if there exist scalars p, q and r such that

$$p\mathbf{u} + q\mathbf{v} + r\mathbf{w} = \mathbf{0}.$$

Similarly, **u**, **v** and **w** are said to be linearly independent if and only if the relation $p\mathbf{u} + q\mathbf{v} + r\mathbf{w} = \mathbf{0}$ implies that $p = q = r = 0$.

Given **u** and **v**, the dot product **u.v** is defined to be

$$\mathbf{u.v} = u_1v_1 + u_2v_2 + u_3v_3.$$

If the two vectors **u** and **v** happen to be perpendicular to each other then

$$\mathbf{u.v} = 0.$$

Given θ is the angle between **u** and **v**, we can also define

$$\mathbf{u.v} = |\mathbf{u}||\mathbf{v}| \cos\theta.$$

If the two vectors **u** and **v** are linearly independent then the cross-product vector $\mathbf{w} = \mathbf{u} \times \mathbf{v}$ is defined as

$$\mathbf{u.w} = u_1w_1 + u_2w_2 + u_3w_3,$$

$$\mathbf{v.w} = v_1w_1 + v_2w_2 + v_3w_3.$$

More specifically, **w** is given as

$$\mathbf{w} = (u_2v_3 - u_3v_2, u_3v_1 - u_1v_3, u_1v_2 - u_2v_1).$$

Note that **w** is orthogonal to **u** and **v**.

If **u** and **v** are parallel to each other then $\mathbf{u} \times \mathbf{v} = 0$.

Moreover, given $\mathbf{u} \neq 0$ and $\mathbf{v} \neq 0$,

$$\|\mathbf{u} \times \mathbf{v}\| = \|\mathbf{u}\|\|\mathbf{v}\| \sin\theta,$$

where $0 \leq \theta \leq \pi$ is the angle between **u** and **v**.

1.2 Lines and Planes in $\mathbf{R}^3$

In $\mathbf{R}^3$, a straight line is defined by any two points. Alternatively one could define a straight line by a point $\mathbf{p}_1$ and a vector $\mathbf{v}$ which defines the direction of the line. Thus, any point $\mathbf{p}$ on the line can then be defined as,

$$\mathbf{p} = \mathbf{p}_1 + s\mathbf{v},$$

where s is a scalar.

Any point (x, y, z) on the line passing through the point $\mathbf{p}_1 = (p_1, p_2, p_2)$ and parallel to the non-zero vector $\mathbf{v} = (v_1, v_2, v_3)$ has the parametric equations of the form

$$x = p_1 + v_1 s,$$
$$y = p_2 + v_2 s,$$
$$z = p_3 + v_3 s.$$

A plane Υ in $\mathbf{R}^3$ can be defined by a point $\mathbf{p}_1$ through which Υ passes, whereby there exist a line through $\mathbf{p}_1$ which is orthogonal to Υ. Another way of defining a plane is by means of a point $\mathbf{p}_1$ on Υ and a vector $\mathbf{n}$ which is orthogonal to the plane Υ. Thus, given that $\mathbf{p}_1 = (x_1, y_1, z_1)$ and $\mathbf{n} = (n_1, n_2, n_3)$, any point $\mathbf{p} = (x, y, z)$ on the plane can be written as

$$n_1 x + n_2 y + n_3 z = c,$$

where $c = n_1 x_1 + n_2 x_2 + n_3 x_3$.

One can also note that given any non-collinear three points we can determine a unique plane passing through them. In order to compute such a plane, one can first compute the displacement vector $\mathbf{u}$ and $\mathbf{v}$ between any two of the points. Then the normal vector $\mathbf{n}$ can be determined as $\mathbf{n} = \mathbf{u} \times \mathbf{v}$. Now, we can use the third point given and the normal vector $\mathbf{n}$ to obtain the equation of the plane.

1.3 Matrix Algebra and Solving Linear Systems

An $m \times n$ matrix (m rows, n columns) is represented as

$$A = \begin{bmatrix} a_{11} & a_{12} & \cdots & a_{1n} \\ a_{21} & a_{22} & \cdots & a_{2n} \\ \vdots & \vdots & \ddots & \vdots \\ a_{m1} & a_{m2} & \cdots & a_{mn} \end{bmatrix}.$$

This can be written as $A = [a_{ij}]$, where the entry in the ith row and jth column is a_{ij}.

1.3.1 Properties of Matrices

Transpose   The matrix $A^T = [a_{ji}]$ formed by interchanging the rows and columns of A is called the transpose of A.

Trace   For an $n \times n$ matrix (which is also known as a square matrix A), the sum of the leading diagonal elements $(a_{11}, a_{22}, \ldots, a_{nn})$ is $\sum_{k=1}^{n} a_{kk}$. This sum is called the trace of A.

Addition   For two matrices of the same size, their sum is defined by adding (or subtracting) the corresponding entries, that is, if $B = [b_{ij}]$ and $C = [c_{ij}]$, then

$$B + C = [b_{ij} + c_{ij}].$$

Multiplication   To multiply a matrix A by a number c (a scalar), we can multiply each entry of A by c, that is,

$$cA = [ca_{ij}].$$

Given A is an $m \times n$ matrix and B is an $n \times p$ matrix, the product AB is the $m \times p$ matrix whose (i, j)th entry is given by

$$\sum_{k=1}^{n} a_{ik} b_{kj}.$$

Note that an $n \times n$ matrix A with all its main diagonal entries equal to 1 and all other entries equal to 0 is called the identity matrix I, and for this matrix we have $AI = IA = A$.

Inverse   Given a square matrix A, its inverse A^{-1} is defined if A^{-1} satisfies $AA^{-1} = A^{-1}A = I$.

Determinant   Given a 2×2 matrix C, the determinant $\det C = |c_{ij}|$ is defined to be $\det C = ad - bc$, where the entries of C are

$$C = \begin{bmatrix} a & b \\ c & d \end{bmatrix}.$$

In general, if A is $n \times m$ then along row i,

$$\det A = a_{i1}c_{i1}(A) + a_{i2}c_{i2}(A) + \cdots + a_{in}c_{in}(A),$$

and along column j,

$$\det A = a_{1j}c_{1j}(A) + a_{2j}C_{2j}(A) + \cdots + a_{nj}c_{nj}(A).$$

1.3.2 Solving Systems of Linear Equations

Systems of linear equations often arise in geometric design applications. A general system of linear equations can be written as

$$a_{11}x_1 + a_{12}x_2 + \cdots + a_{1n}x_n = b_1,$$
$$a_{21}x_1 + a_{22}x_2 + \cdots + a_{2n}x_n = b_2,$$
$$\vdots \qquad \vdots \qquad \ddots \qquad \vdots \qquad \vdots$$
$$a_{m1}z_1 + a_{m2}x_2 + \cdots + a_{mn}x_n = b_m.$$

One can write the above system in matrix form as

$$\begin{bmatrix} a_{11} & a_{12} & \cdots & a_{1n} \\ a_{21} & a_{22} & \cdots & a_{2n} \\ \vdots & \vdots & \ddots & \ddots \\ a_{m1} & a_{m2} & \cdots & a_{mn} \end{bmatrix} \begin{bmatrix} x_1 \\ x_2 \\ \vdots \\ x_n \end{bmatrix} = \begin{bmatrix} b_1 \\ b_2 \\ \vdots \\ b_m \end{bmatrix}.$$

The above can be further represented in augmented matrix form as

$$\begin{bmatrix} a_{11} & a_{12} & \cdots & a_{1n} & b_1 \\ a_{21} & a_{22} & \cdots & a_{2n} & b_2 \\ \vdots & \vdots & \ddots & \vdots & \vdots \\ a_{m1} & a_{m2} & \cdots & a_{mn} & b_m \end{bmatrix}.$$

There are several ways one can solve the above system [1, 5]. A popular approach to solve such a system is by performing a sequence of elementary row operations on the augmented matrix. In particular, the following row operations can be performed:

- Interchange of two rows
- Multiplication of one row by a non-zero number
- Adding a multiple of one row to a different row

Gaussian elimination is a technique that can be efficiently implemented to solve a given system of linear equations [3].

1.4 Properties of Surfaces

There are several ways a surface in $\mathbf{R}^3$ can be represented. One way to represent a surface is using the equation of the form

$$f(x, y, z) = 0.$$

This is known as an implicit surface representation. For example, if $f(x, y, z) = 0$ is of the form $ax + by + cz = d$ then we obtain the equation of a plane. Similarly, if f is of the form $x^2 + y^2 + z^2 = r^2$ then we obtain the representation of a sphere. Many surfaces can be represented in this form. However, for complex surfaces such as those occurring in the real world, implicit surface representation has limitations. An alternative and more popular way to represent surfaces is to use the parametric form.

1.4.1 Parametric Surface Representation

A parametric surface is a surface in the Euclidean space $\mathbf{R}^3$ which is defined by a parametric equation with two parameters. Parametric representation is the most general way to specify a surface. The simplest type of parametric surfaces are defined by graphs of functions of two variables. For example,

$$z = f(x, y), \qquad \mathbf{g}(x, y) = (x, y, z).$$

In general, if we take the real parameters u and v, then the surface can be defined by the vector-valued function, $\mathbf{r} = \mathbf{r}(u, v)$, where $\mathbf{r}(u, v) = (r_1(u, v), r_2(u, v), r_3(u, v))$ and $u_1 \le u \le u_2, \ v_1 \le v \le v_2$.

Parametric surface representation enables generating all types of surfaces and facilitates efficient mathematical analysis of surface properties [2, 4, 6].

1.4.1.1 Properties of Parametric Surfaces

Given a parametric surface of the form $\mathbf{r}(u, v) = (r_1(u, v), r_2(u, v), r_3(u, v))$, one can define the coordinate vectors for the surface

$$\mathbf{r}_u = \frac{\partial r}{\partial u}, \qquad \mathbf{r}_v = \frac{\partial r}{\partial v}.$$

Tangent Plane Given u and v are defined as above and if u and v are parameterized by t, then $\mathbf{r}(t)$ is a curve on the surface with a velocity vector,

$$\mathbf{r}'(t) = \mathbf{r}_u u' + \mathbf{r}_v v'.$$

Now if one considers a point $\mathbf{p}$ on the line, the vector $\mathbf{r}'(t)$ is simply the linear sum of the vectors $\mathbf{r}_u$ and $\mathbf{r}_v$, which lies on the plane determined by these vectors. This plane is known as the tangent plane.

Unit Normal The unit normal $\mathbf{n}$ at any point on a parametric surface is obtained as

$$n = \frac{\mathbf{r}_u \times \mathbf{r}_v}{|\mathbf{r}_u \times \mathbf{r}_v|}.$$

Surface Area The surface area A of a parametric surface can be calculated by integrating the length of the normal vector $\mathbf{r}_u \times \mathbf{r}_v$ to the surface over an appropriate region R in the (u, v) parametric plane, i.e.

$$A(R) = \iint_R |\mathbf{r}_u \times \mathbf{r}_v| \, du \, dv.$$

The First Fundamental Form Given a parametric surface $\mathbf{r}(u, v)$, if we define the quantities $E = \mathbf{r}_u.\mathbf{r}_u$, $F = \mathbf{r}_u.\mathbf{r}_v$ and $G = \mathbf{r}_v.\mathbf{r}_v$ then the first fundamental form I of the surface is the quadratic expression defined as,

$$I = E\,du^2 + 2F\,du\,dv + G\,dv^2.$$

Note that the surface area can also be expressed in terms of the coefficients of the first fundamental form as

$$A(R) = \iint_R \sqrt{EG - F^2}\,du\,dv.$$

The Second Fundamental Form Given a parametric surface $\mathbf{r}(u, v)$ and its normal vector $\mathbf{n}$, if we define the quantities $L = \mathbf{r}_{uu}.\mathbf{n}$, $M = \mathbf{r}_{uv}.\mathbf{n}$ and $N = \mathbf{r}_{vv}.\mathbf{n}$ then the second fundamental form II of the surface is the quadratic expression defined as

$$II = L\,du^2 + 2M\,du\,dv + N\,dv^2.$$

Gaussian and Mean Curvature The Gaussian curvature K and mean curvature H of the surface can be computed using the coefficients of the first and second fundamental forms of the surface, i.e.

$$K = \frac{LN - M^2}{EG - F^2},$$

$$H = \frac{EN - 2FM + GL}{2(EG - F^2)}.$$

1.5 Summary

In this chapter, we have introduced to the reader the basic mathematical concepts for geometric design. Thus, understanding of vector algebra, matrices, systems of linear equations, and particularly mathematical properties of surfaces are essential for understanding and implementing the concepts of geometric design.

References

1. Anton H (2000) Elementary linear algebra. Wiley, New York
2. Hsiung CC (1981) A first course in differential geometry. Wiley, New York
3. Press WH, Teukolsky SA, Vetterling WT, Flannery BP (1992) Numerical recipes in C. Cambridge University Press, Cambridge
4. Pressley A (2003) Elementary differential geometry. Springer, Berlin
5. Sterling MJ (2009) Linear algebra for dummies. Wiley, New York
6. Struik D (1961) Lectures on classical differential geometry. Addison-Wesley, Reading

Chapter 2
Introduction to Geometric Design

Abstract This chapter provides an introduction to geometric design. It introduces various popular mathematical methods used for shape representation in geometric design. It also discusses the role of interactive design and parametric design to enhance the processes involved in a geometric design problem. Furthermore, this chapter discusses the use of design optimization to carry out automatic design for function.

2.1 Introduction

Geometric design concerns with the mathematical description and analysis of shape. Geometric design draws upon the fields such as algebra, geometry, numerical analysis and computer programming. Let us consider the process involved in the design of a new engineering product. Often such a process starts with a definition of a template shape where the requirements in terms of the product's geometric shape and its functionality are specified. This process then proceeds through a sequence of iterative activities to seek an optimal design. Today, this process of 'automatic design for function' relies on the increased use of computers. Although geometric design based on the extensive use of computers does not automatically provide the solution to a given design problem, it can increase the efficiency of the design process. Thus, the main processes of geometric design involve the efficient description of the geometric shape and the integration of the shape with functional analysis. For this purpose, for geometric design, a mathematical method which can generate complex geometries and can relate to the functionality of the object at an early stage of the design process is desirable.

Over the past 40 years, the use of geometric design methods has grown explosively. Today, virtually all computer-based design tasks commence with the use of Computer Aided Design (CAD) systems to create detailed geometric models. These models serve as the point of departure for diverse analysis tools, such as computational fluid dynamics (CFD), stress analysis, geophysical data exploration, and computational electromagnetics or acoustics. Due to the increase in the power of computer hardware, industries such as those related to aerospace, automotive and electronics make more and more integrated use of CAD and analysis. This provides a 'virtual laboratory' for assessing performance characteristics (such as structural

H. Ugail, *Partial Differential Equations for Geometric Design*,
DOI 10.1007/978-0-85729-784-6_2, © Springer-Verlag London Limited 2011

strength or aerodynamic drag) that otherwise would require expensive and time-consuming physical experimentation.

As mentioned above as part of the process of geometric design, the functional properties of the object being created are analyzed by solving the field equations governing the physical process(es) under consideration. One major difficulty encountered here is the linking of complicated surface geometry to analysis [1, 2].

With the assimilation of CAD systems and analysis tools in the major industrial processes, an integrated approach is certainly desirable. However, the need for a systematic way of considering the relationship between geometry and the functional aspects of the geometric model becomes paramount [3].

A mathematical method which can generate complex geometries and can relate to the functionality of the object at an early stage of the design process is desirable. Furthermore, it will be an added advantage if such a method can create a parameterized representation of the object as the variation of the parameters that will then enable the creation of alternative descriptions of the geometry in question while maintaining the functional relations. Such alternative models are necessary for design optimization where the best available design candidate is chosen out of a possible range of designs.

2.2 Mathematical Methods for Shape Representation in Geometric Design

In geometric design, it is common practice to represent geometry of complex shapes in terms of polynomial functions of two parameters. The nature of the surface obtained using such polynomial-based methods usually depends on the type of polynomial chosen. Examples of such surfaces are Bézier surfaces [4], B-splines [5], rational B-splines [6] and non-uniform rational B-splines (NURBS) [7, 8].

A typical bicubic patch in its parametric form can be described as

$$\mathbf{p}(u, v) = \sum_{i=0}^{3} \sum_{j=0}^{3} \mathbf{a}_{ij} u^i v^j, \quad u, v \in [0, 1], \tag{2.1}$$

where $\mathbf{p}$ is a vector of the Cartesian coordinates of points on the surface, u and v are parametric coordinates, and the $\mathbf{a}_{ij}$ are vector coefficients that determine the shape of the surface patch.

The bicubic patch was first introduced in 1963 by Ferguson [9], where the coefficients $\mathbf{a}_{ij}$ in Eq. (2.1) can be expressed in terms of the vectors $\mathbf{p}$, $\mathbf{p}_u$, $\mathbf{p}_v$ and $\mathbf{p}_{uv}$ at the four corner points of the surface patch. The terms $\mathbf{p}_u$ and $\mathbf{p}_v$ are taken to be tangents to the surface in each parametric direction and $\mathbf{p}_{uv}$ is termed the twist vector. The effect of the twist vectors is not intuitively obvious and in his original work Ferguson set them to zero. Ferguson patches are thus expressed in terms of positional and derivative information at the patch corners and can be considered to be obtained from Hermite polynomial interpolation between the corner points. A Ferguson patch can be interpreted as a specific form of the more general Coons patch [10]. The main difference between a Coons patch and a Ferguson patch is that the former is obtained by interpolation between the boundaries of arbitrary form while the latter can be obtained by using parametric cubic boundary curves.

Another common type of surface patch is the so-called Bézier patch,

$$\mathbf{p}(u, v) = \sum_{i=0}^{m} \sum_{j=0}^{n} \mathbf{p}_{ij} B_{i,m}(u) B_{j,n}(v), \quad u, v \in [0, 1], \tag{2.2}$$

where the $\mathbf{p}_{ij}$ are the Cartesian coordinates of the vertices or the 'control points' which form a characteristic polyhedron with an $(m + 1) \times (n + 1)$ rectangular array of points. The $B_{i,m}(u)$ and $B_{j,n}(v)$ are known as Bernstein basis functions and are defined by

$$B_{i,m}(u) = \frac{m!}{i!(m - i)!} u^i (1 - u)^{m-i}, \tag{2.3}$$

and similarly for $B_{j,n}(v)$. A Bézier surface approximates the characteristic polyhedron, and interactive surface design is achieved by moving the control points. The bicubic Bézier patch, for which $m, n = 3$, is essentially a reformulation of the Ferguson patch [9].

An alternative to the Bézier patch is the B-spline surface, which is also defined in terms of the characteristic polyhedron [11]. B-spline surface patches permit the use of more control points in the characteristic polyhedron whilst retaining low order basis functions. They are obtained by replacing the Bernstein basis functions $B_{i,m}(u)$ and $B_{j,n}(v)$ in Eq. (2.2) by the B-spline basis functions $N_{i,k}(u)$ and $N_{j,l}(v)$. The B-spline basis functions are defined recursively by the following formulae:

$$N_{i,1}(u) = \begin{cases} 1 & \text{if } t_i \le u, v \le t_{i+1}, \\ 0 & \text{otherwise,} \end{cases} \tag{2.4}$$

$$N_{i,k}(u) = \frac{(u - t_i)}{(t_{i+k-1} + t_i)} N_{i,k-1}(u) + \frac{(t_{i+k} - u)}{(t_{i+k} - t_{i+1})} N_{i+1,k-1}(u), \tag{2.5}$$

and similarly $N_{j,l}(v)$. The parameters k and l control the degrees $(k - 1)$ and $(l - 1)$ of the resulting polynomials in u and v, and thus also control the continuity of these curves. The t_i and t_j are called knot values and they relate the parametric variables u and v to the $\mathbf{p}_{ij}$ control points. The functions $N_{i,1}(u)$ and $N_{j,1}(v)$ switch between the values 1 and 0 depending on the values of u and v. These B-spline basis functions are non-zero only over a given finite interval and enable the effect of a control point on the surface shape to be localized. Another advantage of the B-spline formulation is its ability to preserve arbitrarily high degrees of continuity over the complex surface patch. These characteristics make the B-spline surfaces popular for use in an interactive modeling environment.

The B-spline formulation was extended to non-uniform rational B-splines (NURBS) by Versprille [12]. The term rational refers to the ratio of the polynomials that characterizes this approach, i.e. a NURBS surface is the ratio of the two B-spline functions [13, 14].

A NURBS surface is defined as

$$S(u, v) = \frac{\sum_{i=0}^{m} \sum_{j=0}^{n} \underline{p}_{i,j} w_{i,j} B_{i,k}(u) B_{j,l}(v)}{\sum_{i=0}^{m} \sum_{j=0}^{n} w_{i,j} B_{i,k}(u) B_{j,l}(v)}. \tag{2.6}$$

The surface has $(m + 1) \times (n + 1)$ control points $\underline{p}_{i,j}$ and weights $w_{i,j}$. Assuming the degrees of basis functions along u and v axes to be $k - 1$ and $l - 1$ respectively, the number of knots is $(m + k + 1) \times (m + l + 1)$. The non-decreasing knot sequence is $t_0 \leq t_1 \leq \cdots \leq t_{m+k}$ along the u direction and $s_0 \leq s_1 \leq \cdots \leq s_{n+1}$ along the v direction with the parameter domain in the range: $t_{k-1} \leq u \leq t_{m+1}$ and $s_{l-1} \leq v \leq s_{n+1}$. If the knots have multiplicity k and l in the u and v directions, respectively, the surface patch will interpolate the four corners of the boundary control points.

Like the rational B-splines NURBS are infinitely smooth in the interior of the knot span provided the denominator is not zero; and at a given knot NURBS are at least C^{k-1-r} continuous with knot multiplicity r, which enable them to satisfy different smoothness requirements. NURBS also share properties such as the 'convex hull' property, 'local support' and invariance under standard geometric transformations [14]. Additionally, the weights $w_{i,j}$ act as extra degrees of freedom influencing the local shape, i.e. if a particular weight is set to zero, then the corresponding rational basis function is also zero, and its control point does not effect the NURBS shape. The spline is attracted towards a control point more if the corresponding weight is increased and less if the weight is decreased. Moreover, NURBS also form a common mathematical framework for both implicit and parametric forms, i.e. in principle they can represent analytic functions such as conics and quadratics as well as free-form shapes.

The spline based definition for curves and surfaces forms the basis for many of today's geometric design systems. However, to create a given object, the chosen geometric design system may use a variety of analytic descriptions for curves and surfaces, or the system may use a combination of analytic forms and spline based functions to perform operations, such as union, difference and intersection [15]. Furthermore, some geometric design systems use variational modeling schemes in which the basic spline functions are manipulated using physically based relations, such as force and energy [16].

2.2.1 Schemes for Geometry Model Representation

Various geometry representation techniques have been developed to represent two-dimensional or three-dimensional geometric shapes. Popular representation techniques include: Boundary Representation (B-Rep), Constructive Solid Geometry (CSG), feature based representations and variational geometry.

B-Rep Approach   In a B-Rep approach, a shape is represented by the boundary information such as faces, edges and vertices, i.e. B-Rep represents geometry in terms of boundaries and topological relations.

CSG Approach   The CSG approach models geometric shapes using a set of 'primitives' such as cubes, cylinders or prisms. Complex shapes are built from the primitives through a set of Boolean operations (e.g. union, difference and intersection). Most CSG systems in use today offer quite a variety of primitive solids, ranging from various types of spheres and ellipses, boxes and cones, and solids defined by swept or extruded curves. The CSG modeling approach has several inherent limitations of which the most notable limitation is the non-uniqueness of a CSG representation. This non-uniqueness of representations makes recognition of shapes from their CSG representation extremely difficult.

Feature Based Approach   In the feature based representation, a part is built from a set of feature 'primitives'. Examples of features include holes, slots and ribs. A feature based design approach allows a user to use features stored in a feature library. It provides a means for building a complete CAD database with the features right from the start of the design. However, this approach suffers from the difficulty of there being a limited number of available feature primitives. It is difficult to satisfy various design needs, and in the event that the features interact with one another, new features may arise that can cause complication with the analysis process. Feature based design allows a designer to bridge the gap between units of the designer's perception of forms and data in geometric models. In this scheme of representation, shapes are described in the way the designer understands them [17].

Variational Approach   The concept of using variational geometry in geometric design started as early as 1981. Instead of defining a geometric model with respect to a set of characteristic points in $\mathbf{R}^3$, dimensions are treated as constraints limiting the permissible locations of these points. Many schemes for variational design have been suggested, e.g. [18–21].

Many of these schemes use a physical analogy in which a chosen functional is used to minimize the elastic energy satisfying certain interpolation constraints imposed on the mechanism by which the surface is created. The method of Partial Differential Equations [22–24] which is discussed in this book falls into this category.

It is important to note that the modeling scheme we choose forms an integral part of the geometric design process. To be useful within a given application area, the range of shapes that can be represented by a given scheme should be adequate. Moreover, the scheme should be user friendly, i.e. the model representation scheme should be well suited for 'interactive design'.

2.3   Enhancing Geometric Design Using Interactive and Parametric Design

In the early days of geometric design, design applications were carried out in 'batch', i.e. a complete task (or job) was first defined by the user and then submitted to the computer. The computer processed the complete job without further interaction from the user and then produced an output.

Most of the existing geometric design systems, if not all, make heavy use of interactive graphics techniques rather than batch techniques. Thus, the user can interact with the computer via input devices such as the mouse and keyboard.

2.3.1 Techniques for Interactive Design

As discussed above, in geometric design, it is common practice to describe the geometric models by means of spline based methods. There exist many techniques for interactive design using such methods. Perhaps the most basic case is the use of Bézier patches in which the displacement of a control point results in the change in the shape. This technique has also been applied to B-splines to control the shape of the surface patch locally. Applying the above technique to an isolated control point frequently leads to results with an unpredictable effect on the resulting shape of the surface. Designers usually face the questions of choosing which control points to move in which direction [25]. In principle, it is possible to produce large-scale changes to the shape of the surface by moving more than one control point. However, such interactive manipulations often result in undesired bumps or wiggles within the surface patch.

As far as interactive design methods using NURBS are concerned, an initial surface is created via specification of a control polygon. The initial shape is then refined into the final desired shape through interactive adjustments of control points and weights and possibly addition and deletion of knots. The knot insertion algorithm [26], the control point insertion algorithm [27] are all complementary elements for interactive shape refinements. However, such refinement processes are often considered to be tedious and very unpredictable [28]. For example, to adjust the shape of a surface should a designer move a control point, or change a weight?

Despite the recent advent of sophisticated devices for 3D interaction, the above mentioned techniques for interactive surface design and manipulation can be difficult for a designer to use effectively. To overcome this problem, techniques which allow 'physically based' manipulation have been introduced. Many authors have suggested the use of 'constraint based interface', where some of the design parameters have some form of physical relevance.

For example, Terzopoulos and Watkin describe simple interactive sculpting using viscoelastic and plastic models [29]. Celniker and Gossard [30] describe an interesting prototype system for interactive free-form design based on the finite-element optimization of energy functionals. Thingvold and Cohen [31] proposed a deformable B-spline whose control points are mass points connected by elastic springs and hinges. Celniker and Welch [30] investigated deformable B-splines with linear constraints. Furthermore, for design using NURBS, free-form deformable models were introduced by Terzopoulos et al. [32]. Such models were further developed by Pentland and Williams [33], Platt and Barr [34]. A similar technique for real time design using deformations is discussed by Borrel and Rappoport [35].

Terzopoulos and Qin, on the other hand, describe a model for interactive design in which they use a generalized form of NURBS called Dynamic NURBS or

D-NURBS. The D-NURBS model is governed by dynamic differential equations which, when integrated numerically through time, continuously evolve the control points and weights in response to applied forces [36].

Unlike models based on the direct manipulation of surfaces, the behavior of deformable models are governed by 'physical' laws. The result is that such models respond to the user interactions in a natural and somewhat predictable way. Many existing geometric design systems use these techniques. However, as far as surface manipulations in such systems are concerned, the initial surface is often provided as a pre-defined geometry model obtained from scan-data, for example, on which only small scale manipulations are allowed to be carried out.

An important point to note in the existing mathematical models which allow interactive manipulations of surfaces is the large number of design parameters often involved. In the development of effective mathematical models, for the purpose of interactive design, much effort has been put into trying to reduce the number of design parameters associated with the chosen model. Moreover, much effort has been concentrated towards choosing design parameters with a readily apparent physical meaning. Thus, the use of 'parametric design' has recently been very popular.

2.3.2 Parametric Design

One of the requirements for geometric design systems is the ability to parameterize the shape of objects. In parametric design, the basic approach is to develop a generic description of an object or class of objects, in which the shape is controlled by the values of a set of design variables or parameters. A new design, created for a particular application, is obtained from this generic template by selecting particular values for the design parameters so that the item has properties suited to that application.

The design of a wide range of manufactured products conforms to this general pattern, ranging from engine components to such objects as aircrafts. If a product's geometry is composed of standard geometric 'constructs' such as circles, ellipses, cylinders, etc., then the parameterizations of its shape is relatively straightforward. However, for most products at least some parts of their shape are composed of freeform surfaces which, although they may constitute a small fraction of the total surface area, can, nevertheless, be very important functionally.

Many geometric design systems can handle the parameterizations of 'standard' shapes, though for objects with complicated shapes, commercial systems often fail, owing to the inability of the geometry modeling package to parameterize such shapes. Particular problem areas include the generation of surfaces which do not conform to standard, limited descriptions.

The inherent problem with the mathematical models which are used to describe the geometry of a given model is the nature of their complexity. This is particularly problematic when design has to be carried out from 'scratch' in an interactive environment. Thus, a mathematical model which can model and parameterize the geometry in terms of small set of shape parameters and, also, enable a quick interaction with the geometry is desirable.

2.4 Use of Optimization Techniques in Geometric Design

A typical demand in a practical geometric design task may be to minimize or maximize an objective function without violating a set of constraints. In order to improve a design by applying methods of computational optimization, it is necessary to express the design objective and constraints of the optimization problem by an appropriate mathematical formulation. A general formulation of the optimization problem can be written as

$$\min\{f(\mathbf{x}) \mid \mathbf{x}_l \leq \mathbf{x} \leq \mathbf{x}_u; \ \mathbf{g}(\mathbf{x}) = \mathbf{0}; \ \mathbf{h}(\mathbf{x}) \leq \mathbf{0}\}, \quad \mathbf{x} \in \mathfrak{R}^n, \tag{2.7}$$

with

f	the objective function;
$\mathbf{x}$	vector of n design variables;
$\mathbf{g}$	vector of p equality constraints;
$\mathbf{h}$	vector of q inequality constraints;
$\mathbf{x}_l$ and $\mathbf{x}_u$	lower and upper bounds for the design variables.

The design variables and the constraints form the feasible design space

$$\mathbf{x} \in R^n \mid \mathbf{x}_l \leq \mathbf{x} \leq \mathbf{x}_u; \ \mathbf{g}(\mathbf{x}) = 0; \ \mathbf{h}(\mathbf{x}) \leq 0, \tag{2.8}$$

which describes the design space.

Coming up with appropriate formulations of the design objective and constraints of the optimization problem in a geometric design application is not always a trivial task. For example, due to the complex nature of most engineering problems, the choice of the right objective function requires experience and the fundamental understanding of the design objectives. Furthermore, not all constraints can be easily formulated in a mathematically correct way for optimization [37].

There exist a wide variety of methods for numerical optimization. The choice of a particular method is problem specific and involves considerations such as the computational cost of evaluating the function to be optimized and also the behavior of the function within the design space. Generally, these methods can be divided into two categories, i.e. those that only require the evaluation of the objective function and those that require the evaluation of the objective function and its derivatives with respect to the design parameters. During the process of optimization, most of the computational effort is spent on evaluating the objective function rather than in the optimization routine itself. Therefore, it is desirable to use a design method which minimizes the number of design variables and therefore requires as few function evaluations as possible.

Generally, the optimization process requires a search to be made in the parameter space in order to find the minimum value of the objective function. (Note that, without loss of generality, we can consider minimization problems, since maximizing a function f is equivalent to minimizing $-f$.) Particular algorithms which are used for minimization in numerical analysis include the downhill simplex algorithm due to Nelder and Mead [38] and Powell's direction set algorithm [39].

Nearly all these gradient-based methods have the common feature that they perform a series of local minimizations in which the objective function f is minimized along a straight line in the parameter space. These methods are iterative and at each successive iteration they give a vector $\mathbf{x}^k = (x_1{}^k, x_2{}^k, \ldots, x_n^k)$ of the n independent design variables which is computed from the previous iterations using the expression

$$\mathbf{x}^{k+1} = \mathbf{x}^k + \alpha^k \mathbf{s}^k. \tag{2.9}$$

Here $\mathbf{s}^k$ is a direction of search and α^k is a scalar that minimizes the one-dimensional function $F(\alpha) \equiv f(\mathbf{x}^k + \alpha^k \mathbf{s}^k)$. Thus, given a starting point, the algorithm moves in a series of steps through points in the parameter space, giving a lower value of the objective function than previously, until it finds a lowest possible local value of the objective function. An important point to note regarding this type of methods is that they find local minima. Thus, if a global minimum is required, multiple searches by such methods have to be performed with different starting points.

Very often it is the case that the design space considered contains many local minima, and it becomes extremely difficult to search for a global minimum using local minimization methods. An alternative method, which is considered in this work, is a global optimization method which uses a stochastic process known as Simulated Annealing [40, 41]. This method probabilistically searches in every region of the design space and therefore converges to a global minimum although not necessarily in a finite time.

As mentioned before, the particular algorithm used for numerical optimization must take into account the computing time needed to evaluate the objective function. Each function evaluation must be performed and thus may be very costly in terms of computing time.

Various approaches have been taken to perform the actual optimization. A typical approach [42] is to consider the optimization in terms of successive linear programming problems. The constraints and the objective function are linearized about the current design variable values and this simplified problem is solved. The result is taken as the new design variable values and the process is repeated until no further improvements can be made. This method has the advantage of making use of the efficient linear programming algorithms that are available. Alternatively, the search algorithm can be based on the design sensitivities to solve the full non-linear problem, again in an iterative manner. In most design optimization, it is common that the design variables are effectively taken as the Cartesian coordinates of points that boundary curves of a particular form were required to pass through. However, a different approach to this was taken by Kristensen and Madsen [44] who described the boundary as a weighted sum of certain specified functions, the weights being taken as the design variables.

It should be noted that the most important aspect of shape optimization is the choice of the design variables to be used and how the boundary shape is parameterized in terms of these design variables. Choosing too many variables will considerably complicate the design problem with severe implications on the computational time required, and having too few variables may result in only trivial solutions being obtained [43, 45]. It is therefore a basic requirement that a wide range of boundary shapes (which can be defined by a relatively small number of parameters) are accessible to the method of optimization used.

2.5 Summary

In this chapter, we have given an introduction to the geometric design. The discussions have been centered around some of the popular methods for geometry representation such as splines. The key points to consider when developing a geometric design system are the ability to represent a given object in an efficient way, the ability then to create alterative designs using parametric representation, and the ability to generate an optimal design by means of careful consideration of alterative designs in a consistent fashion via the use of numerical optimization techniques.

References

1. Farouki RT (1999) Closing the gap between CAD model and downstream application. SIAM News 32:5. http://www.siam.org/siamnews/
2. Lee W (1999) Principles of CAD/CAM/CAE systems. Addison-Wesley, Reading
3. Shapiro V, Voelcker H (1989) On the role of geometry in mechanical design. Res Eng Des 1:63–73
4. Bézier P (1986) The mathematical basis of UNISURF CAD system. Butterworths, London
5. Woodward CD (1987) Blends in geometric modelling. In: Martin RR (ed) Mathematical methods of surfaces II. Oxford University Press, London, pp 255–297
6. Tiller W (1983) Rational B-splines for curve and surface representation. IEEE Comput Graph Appl 3:61–69. doi:10.1016/0010-4485(87)90234-X
7. Piegl L, Tiller W (1987) Curve and surface constructions using rational B-splines. Comput Aided Des 19:485–498. doi:10.1016/0010-4485(87)90234-X
8. Schumaker LL (1981) Spline functions: basic theory. Wiley, New York
9. Faux DI, Pratt MJ (1979) Computational geometry for design and manufacture. Ellis Horwood, Chichester
10. Coons SA (1994) Surfaces for computer-aided design of space forms. Project MAC, Report MAC-TR-41, Massachusetts Institute of Technology
11. Mortenson ME (1985) Geometric modelling. Wiley, New York
12. Vasprille KJ (1975) Computer-aided design applications of the rational B-spline approximation form. PhD thesis, Syracuse University, Syracuse, New York
13. Piegl L (1991) On NURBS: a survey. IEEE Comput Graph Appl 11(1):55–71. doi:10.1109/38.67702
14. Farin G (1990) Curves and surfaces for computer-aided design: a practical guide, 2nd edn. Academic Press, New York
15. Hsu W, Woon YI (1998) Current research in the conceptual design of mechanical products. Comput Aided Des 30(5):377–389. doi:10.1016/S0010-4485(97)00101-2
16. Nowacki H, Dingyuan L, Xinmin L (1989) Mesh fairing GC surface generation method. In: Straßer W, Seidel HP (eds) Theory and practise of geometric modelling. Springer, Berlin, pp 93–103
17. Nakajima N, Gossard D (1982) Basic study in feature descriptor. MIT CAD Technical Report, Massachusetts Institute of Technology, Cambridge

18. Nowacki H, Rees D (1983) Design and fairing of ship surfaces. In: Barnhill RE, Boehm W (eds) Surfaces in CAGD. North-Holland, Amsterdam, pp 121–134

19. Hagen H, Schulze G (1990) Variational principles in curve and surface design. In: Hagen H, Roller D (eds) Geometric modelling. Springer, Berlin, pp 161–184

20. Kallay M (1993) Constrained optimisation in surface design. In: Falcidieno B, Kunii TC (eds) Modelling in computer graphics. Springer, Berlin, pp 85–93

21. Brunnett G, Wendt J (1998) Elastic splines with tension control. In: Dæhlen M, Lyche T, Schumaker LL (eds) Mathematical methods for curves and surfaces II. Vanderbilt University Press, Nashville, pp 33–40

22. Bloor MIG, Wilson MJ (1989) Generating blend surfaces using partial differential equations. Comput Aided Des 21(3):33–39. doi:10.1016/0010-4485(89)90071-7

23. Bloor MIG, Wilson MJ (1989) Blend design as a boundary-value problem. In: Straßer W, Seidel HP (eds) Theory and practise of geometric modelling. Springer, Berlin, pp 221–234

24. Ugail H, Bloor MIG, Wilson MJ (1999) Techniques for interactive design using the PDE method. ACM Trans Graph 18(2):195–212. doi:10.1145/318009.318078

25. Farin G, Sapidis N (1989) Curvature and the fairness of curves and surfaces. IEEE Comput Graph Appl 3:25–57. doi:10.1109/38.19051

26. Boehm W (1980) Inserting new knots into B-spline curves. Comput Aided Des 12:199–201. doi:10.1016/0010-4485(80)90154-2

27. Léon JC (1991) Modélisation des courbes et des surfaces pour la CFAO. Hermés, Paris

28. Rappoport A, Helor Y, Werman M (1994) Interactive design of smooth objects with probabilistic point constraints. ACM Trans Graph 13:156–176. doi:10.1145/176579.176582

29. Terzopoulos D, Witkin A (1988) Physically based models with rigid and deformable components. IEEE Comput Graph Appl 8(6):41–51. doi:10.1109/38.20317

30. Celniker G, Gossard D (1991) Deformable curve and surface finite-elements for free-form shape design. Comput Graph 25(4):257–266. doi:10.1145/122718.122746

31. Thingvold JA, Cohen E (1990) Physical modelling with B-spline surfaces for interactive design and animation. Comput Graph 24(2):129–137. doi:10.1145/91385.91430

32. Terzopoulos D, Platt J, Barr A, Fleischer K (1987) Elastically deformable models. Comput Graph 21(4):205–214. doi:10.1145/37401.37427

33. Pentland A, Williams J (1989) Good vibrations: modal dynamics for graphics and animation. Comput Graph 23(3):215–222. doi:10.1145/74333.74355

34. Platt J, Barr A (1988) Constraints methods for flexible models. Comput Graph 22(4):279–288. doi:10.1145/54852.378524

35. Borrel P, Rappoport A (1994) Simple constrained deformations for geometric modelling and interactive design. ACM Trans Graph 13:137–155. doi:10.1145/176579.176581

36. Terzopoulos D, Qin H (1994) Dynamic NURBS with geometric constraints for interactive sculpting. ACM Trans Graph 13:103–136. doi:10.1145/176579.176580

37. Cohen MZ (1994) Theory and practise of structural optimisation. Struct Multidiscip Optim 7:20–31. doi:10.1007/BF01742500

38. Nelder JA, Mead R (1965) A simplex method for function minimisation. Comput J 7:308–313. doi:10.1093/comjnl/7.4.308

39. Press WH, Teukolsky SA, Vetterling WT, Flannery BP (1992) Numerical recipes in C. Cambridge University Press, Cambridge

40. Kirkpatrick S, Gellat D Jr, Vecchi MP (1983) Optimisation by simulated annealing. Science 220(4598):671–680. doi:10.1126/science.220.4598.671

41. Vanderbilt D, Louie SG (1984) A Monte Carlo simulated annealing approach to optimisation over continuous variables. J Comput Phys 56:259–271. doi:10.1023/A:1004680806815

42. Zienkiewicz OCZ, Campbell JS (1973) Shape optimisation and sequential linear programming. In: Gallagher RH, Zienkiewicz OCZ (eds) Optimal structural design. Wiley, London, pp 109–126

43. Raphael TH, Ramana GV (1986) Structural shape optimization: a survey. Comput Methods Appl Mech Eng 57(1):91–106. doi:10.1016/0045-7825(86)90072-1

44. Kristensen ES, Madsen NF (1976) On the optimal shape of fillets in plates subject to multiple in-plane loading cases. Int J Numer Methods Eng 10:1007–1019. doi:10.1002/nme.1620100504
45. Imam MH (1982) Three-dimensional shape optimisation. Int J Numer Methods Eng 18:661–673. doi:10.1002/nme.1620180504

Chapter 3
Introduction to Partial Differential Equations

Abstract This chapter provides an introduction to partial differential equations (PDEs) with the aim of introducing the reader with the mathematical concepts that are used in further chapters. The chapter first introduces the general concept of PDEs and discusses various types of PDEs. Special emphasis is given to elliptic PDEs since this type of equations form the basis for the development of geometric design techniques throughout this book.

3.1 Definition of a PDE

In simple terms, one can describe a PDE as a mathematical tool which can be used to describe a given physical phenomena. This description is given in the form of a mathematical relation between different rates of change of the phenomena in question with respect to different variables, e.g. the physical coordinates and time. Thus, many physical problems in the real world can be mathematically described by some form of a PDE, e.g. physical phenomena such as heat transfer, ripple propagation in a pond, certain problems in economics and finance.

Mathematically speaking, the rate of change of one quantity with respect to another is known as a derivative, and in particular these rates are known as partial derivatives when the function that is being differentiated depends on two or more variables. For instance, assume that a function F depends on x, y and t, that is, $F(x, y, t)$ where $0 \leq x \leq 1$, $0 \leq y \leq 1$ and $t \geq 0$. Thus, the rate of change of $F(x, y, t)$ with respect to the variable x is denoted by the following notation

$$\frac{\partial F}{\partial x},$$

(3.1)

and it represents the partial derivative of F with respect to x. If we assume that the derivative $\frac{\partial F}{\partial x}$ has to be differentiated again, but this time with respect to y then this is written as

$$\frac{\partial^2 F}{\partial y \partial x}.$$

(3.2)

Now if one assumes that a given physical phenomenon is mathematically modeled by F and it is governed by a relation establishing that the sum of the partial derivative of F with respect to x and the second partial derivative of F with respect with

H. Ugail, *Partial Differential Equations for Geometric Design*,
DOI 10.1007/978-0-85729-784-6_3, © Springer-Verlag London Limited 2011

x and y has to be equal to a given function $G(x, y, t)$, then all this can be efficiently represented in a PDE such as

$$\frac{\partial F}{\partial x} + \frac{\partial^2 F}{\partial y \partial x} = G(x, y, t). \tag{3.3}$$

In other words, PDEs are mathematical relations of partial derivatives of a given function. As the reader may expect, the task of translating a given phenomenon into the relevant mathematical language is not necessarily an easy task and may involve physical experimentation as well as intuition. One should note that Eq. (3.3) has no known physical meaning and it was simply posed for illustration purposes.

One of the most crucial parts of the study of PDEs lies in identifying their solutions. Sometimes it is not an easy task and hence specialized areas of mathematics are dedicated to this. Before discussing solution methods for PDEs, it is necessary to provide some examples of PDEs and how we classify them.

3.1.1 Examples of PDEs

PDEs are used in almost all scientific disciplines, and therefore they are often named after the phenomenon they describe or the person who related such an equation to a particular phenomenon and found its solution. Below the reader will find some examples of PDEs [7, 8, 11, 13].

- *Heat equation*

  The distribution of heat in a region or an object over a period of time can be described by

  $$\frac{\partial Q}{\partial t} = k \left(\frac{\partial^2 Q}{\partial x^2} + \frac{\partial^2 Q}{\partial y^2} + \frac{\partial^2 Q}{\partial z^2} \right), \tag{3.4}$$

  where $Q = Q(x, y, z, t)$ is a function describing the temperature at any given position (x, y, z) of the object at a given time t. Here k represents the thermal conductivity of the object or region where heat is being distributed, and it is usually a constant provided the heat is to be transferred through a homogeneous isotropic medium. The heat equation is a particular case of the diffusion equation which can be used to model a number of diffusion problems.

- *Wave equation*

  Light, sound, water and electromagnetic currents are propagated in the form of waves. The equation, in a three-dimensional form, describing how such waves are propagated is given by

  $$\frac{\partial^2 \Psi}{\partial x^2} + \frac{\partial^2 \Psi}{\partial y^2} + \frac{\partial^2 \Psi}{\partial z^2} = \frac{1}{v^2} \frac{\partial^2 \Psi}{\partial t^2}, \tag{3.5}$$

  where v represents the velocity at which the wave propagates through space and Ψ is a function of x, y, z and t.

- *Poisson's equation*

    This equation is commonly used in electrostatics in order to describe an electrostatic field. It can be also used to describe other physical phenomena in areas such as theoretical physics and engineering. The general form of Poisson's equation is given as

$$\frac{\partial^2 \phi}{\partial x^2} + \frac{\partial^2 \phi}{\partial y^2} + \frac{\partial^2 \phi}{\partial z^2} = f, \tag{3.6}$$

where $\phi(x, y, z)$ denotes the potential field and $f(x, y, z)$ is a given function. A particular case in which this equation is employed in electrostatics is

$$f(x, y, z) = 4\pi\rho,$$

where ρ represents the charge density of the medium. Another particular case of this equation considers

$$f(x, y, z) = 0.$$

This case corresponds to the Laplace equation.
- *Black–Scholes equation*

    This equation is commonly used in mathematical finance to predict the value of a given stock and is given as

$$\frac{\partial f}{\partial t} + rS\frac{\partial f}{\partial S} + \sigma^2 S^2 \frac{\partial^2 f}{\partial S^2} = rf, \tag{3.7}$$

where $f = f(S)$ is a general derivative that is a function of S, the underlying variable r denotes the risk free rate of return and σ is a constant representing the volatility.

Other examples include the Navier–Stokes equations used in fluid mechanics, equilibrium equations that are responsible for describing the stress and strain distributions in solid mechanics, and the Schrödinger equation which can be used to describe non-relativistic quantum mechanics.

3.2 Classification of PDEs

There are several criteria which can be employed to classify PDEs. These mainly depend on the order, the linearity and homogeneity of the PDE. Below, some of the most standard criteria to classify PDEs are explained below.

3.2.1 Order

The order of a PDE is determined by the highest order of all the partial derivatives involved in the equation. For instance, the equation given by

$$\frac{\partial^4 h}{\partial x^4} + \frac{\partial h}{\partial t} = 0 \tag{3.8}$$

is a fourth order partial differential equation whereas the equation determined by

$$\frac{\partial f}{\partial x} + \frac{\partial f}{\partial y} + \frac{\partial f}{\partial z} = 0 \tag{3.9}$$

is of the first order. This is perhaps one of the easiest and simplest criteria to identify and classify a PDE.

3.2.2 Homogeneity

The homogeneity of a partial differential equation depends on the existence of terms depending on the independent variables, i.e., if there are terms in which the independent variables are involved, the equation is said to be non-homogeneous and homogeneous if such terms do not exist. For example, the equation

$$\frac{\partial^4 h}{\partial x^4} + \frac{\partial h}{\partial t} + x = 0 \tag{3.10}$$

is a non-homogeneous partial differential equation whereas the equation determined by

$$\frac{\partial f}{\partial x} + \frac{\partial f}{\partial y} + \frac{\partial f}{\partial z} = 0 \tag{3.11}$$

is homogeneous. This criterion is slightly more difficult to identify when compared to the order of a PDE. However, it is just a matter of writing all the relevant terms that include the function which is being differentiated on one side and leaving the rest of the terms on the other side. If there is a non-zero term on the right hand side of the equation it is described as non-homogeneous.

3.2.3 Linearity

The linearity of a PDE is determined by the order of the derivative function involved to describe the PDE. In other words, the unknown function and its derivatives must appear to the power of one at all times and no products among the unknown function and its derivatives are permitted. Otherwise, the PDE is regarded as non-linear. Examples of linear and non-linear partial differential equations are listed below. The equation

$$\frac{h^3 \partial^4 h}{\partial x^4} + \frac{\partial h}{\partial t} + x = 0 \tag{3.12}$$

is a non-linear partial differential equation whilst

$$\frac{\partial f}{\partial x} + \frac{\partial f}{\partial y} + \frac{\partial f}{\partial z} = 0 \tag{3.13}$$

is linear.

There are other more complicated criteria that can be used to classify PDEs. The most common is the use of a discriminant to categorize PDEs as elliptic, parabolic or hyperbolic, depending on the value of the discriminant. Details on how this classification scheme works are given below.

3.2.4 Use of a Discriminant as a Classification Method

Often in mathematics discriminants are used for determining the nature of the roots associated with a given second order algebraic equation. Therefore, a discriminant could easily tell if both roots are real and different, if there is only one root or if they are complex. Similarly, PDEs can be classified into different types of equations depending on the value of the discriminant. This classification is rather useful since it can guide a user to the solution to a particular PDE. Assume that the general second order partial differential equation in two variables is given by

$$A\frac{\partial^2 F}{\partial x^2} + B\frac{\partial^2 F}{\partial y \partial x} + C\frac{\partial^2 F}{\partial y^2} + \cdots = 0. \tag{3.14}$$

This equation has been written under the assumption that

$$\frac{\partial^2 F}{\partial y \partial x} = \frac{\partial^2 F}{\partial x \partial y}.$$

The discriminant through which Eq. (3.14) is classified is thus

$$B^2 - 4AC.$$

The classification is divided into three major groups, i.e.

- $B^2 - 4AC < 0$

  PDEs fulfilling this condition are regarded as elliptic PDEs. An example of such a PDE is the Laplace equation.
- $B^2 - 4AC = 0$

  Any second order PDE satisfying this condition is classified as a parabolic PDE. The heat equation is an example of a parabolic equation.
- $B^2 - 4AC > 0$

  The remaining condition characterizes hyperbolic partial differential equations. An example of such an equation is given in the form of the wave equation.

This classification of second order PDEs can be further extended to PDEs of higher order. However, for the sake of brevity, this will not be included here. A description of each type of partial differential equations is given below so that the reader can appreciate why this classification has been so useful.

3.2.4.1 Elliptic PDEs

The solution of this type of PDEs is generally given in terms of harmonic functions [1], and are smooth within the domain in which they are solved. Moreover, if the coefficients multiplying the terms involving the unknown function and its derivatives exist then the solution can be found using Fourier transforms.

3.2.4.2 Parabolic PDEs

Parabolic PDEs are typically related to evolution problems such as diffusion of heat through a medium. For this reason, they are also known as evolution equations since they describe how a physical property changes through time across a given domain. Generally, solutions to this type of equations are less stable when compared to elliptic PDEs.

3.2.4.3 Hyperbolic PDEs

A traditional example of a hyperbolic PDE is the wave equation and in its simplest form the solution is given in the form of a traveling wave by

$$f(x,t) = U(x + ct), \tag{3.15}$$

where c is the velocity at which the traveling wave propagates.

3.3 Harmonic, Biharmonic and the Triharmonic Equation

Special emphasis is given to the Harmonic, Biharmonic and Triharmonic equations since they represent the foundations of the surface generation technique that will be discussed in this book. These equations fall into the category of elliptic PDEs and are commonly denoted by

$$\left(\nabla^2\right)^k \chi = 0, \tag{3.16}$$

where ∇^2 represents the Laplace operator and $k \geq 1$. The form of the Laplace operator in three-dimensional Cartesian coordinates is

$$\nabla^2 = \frac{\partial^2 \phi}{\partial x^2} + \frac{\partial^2 \phi}{\partial y^2} + \frac{\partial^2 \phi}{\partial z^2}. \tag{3.17}$$

The reader may have already noticed that in the particular case when k is equal to 1, the equation obtained is the Laplace equation mentioned previously. This equation is called the Harmonic equation since its solution is given by Harmonic functions which are functions with continuous partial second derivatives satisfying Laplace's condition. The Harmonic equation has been extremely useful to describe a number of physical phenomena, particularly those that can be related to potential fields. For this reason this type of PDEs are also used to describe laws of conservation. As for the solution of the PDE, the nature of the boundary conditions required to solve such an equation leads to two different approaches:

- *Dirichlet boundary conditions*
  These boundary conditions specify the value of ϕ on the boundary of the region in which the PDE is solved.
- *Neumann boundary conditions*
  These boundary conditions specify the value of the normal derivative of the function ϕ at the boundary of the domain.

3.3.1 The Biharmonic Equation

The case when $k = 2$ in Eq. (3.16) is known as the Biharmonic equation. This time, its solution is given in terms of functions whose fourth partial derivatives are continuous and satisfy the Biharmonic condition. Examples in which this equation has played an important role describing physical phenomena are the case of Stokes flow in fluid dynamics, where the Biharmonic equation is used to find the stream function describing the flow or in continuum mechanics where it is used to find Airy or Love stress functions to describe plane stress or plane strain problems relating to a displacement function.

3.3.2 The Triharmonic Equation

The particular case when $k = 3$ in Eq. (3.16) is known as the Triharmonic equation and, as the reader may already suspect, its solution is given in terms of functions whose sixth order partial derivatives are continuous. This type of equation is mentioned since it has been explored in geometric design as a surface generation technique providing curvature continuity between two surface patches.

It is worth mentioning that the solution to these three equations are so similar that sometimes they look as if the solution to the Harmonic equation has been extended to find the respective solutions of the other two equations. Further details on the methods for solving this type of PDEs are given below.

3.4 Solution Methods

One should note that the task of finding a solution to a PDE is by no means trivial. Sometimes one can be lucky and find a simple analytic solution to the PDE in question. However, in the vast majority of cases this is not the case. It is noteworthy that entire fields of mathematics are devoted to the analysis of PDEs for developing novel methods for finding solutions to PDEs.

Thus, one can find that there are several analytic techniques capable of providing an exact solution, some other analytic techniques leading to approximate solutions and partial or fully numerical techniques for solving very complex problems defined in terms of PDEs. A list of the most important techniques is given below together with a brief description explaining the basic theory behind them.

3.4.1 Analytic Methods

The analytic methods available for finding the solution of a given PDE include the method of separation of variables and the method of change of variable [3]. These methods are explain below.

- *Separation of variables*

     The method of separation of variables is perhaps the most common when attempting to find the solution to a given PDE. This method is generally used to solve linear PDEs and consists of expressing the unknown function in terms of a product of a series of functions, each of which depends only on one of the independent variables. Then this function is substituted in the PDE to be solved. For instance, consider the Harmonic equation in two dimensions given by

$$\frac{\partial^2 u}{\partial x^2} + \frac{\partial^2 u}{\partial x^2} = 0. \tag{3.18}$$

Thus, the method of separation of variables establishes that

$$u(x, y) = X(x)Y(y).$$

After working out the necessary algebra, we can write the equation as

$$\frac{X''}{X} = -\frac{Y''}{Y},$$

where X'' and Y'' denote the second derivative of X and Y, respectively. Note that we have now grouped all the terms depending on x on one side and those depending on y on the other side of the equation. Therefore, the equation has a solution if and only if each of the sides is equal to the same constant, leading to

$$X'' = -k^2 X \quad \text{and} \quad Y'' = k^2 Y.$$

Thus, the solution to Eq. (3.18) is given by,

$$u(x, y) = \left(A \exp(kx) + B \exp(-kx)\right) + \left(C \cos(ky) + D \sin(ky)\right),$$

where A, B, C and D are constants whose value will depend on the particular boundary conditions.

- *Change of variable*

     Solutions to some PDEs also can be obtained by performing a suitable change of variable whereby the original PDE is written in a simple form which is easier to solve. For instance, with a suitable change of variable the Black–Scholes equation described earlier can be reduced to the heat equation.

There are some other analytic methods for solving PDEs such as the method of characteristics and the use of superposition principle. For the sake of brevity, these methods will not be explained here. In addition to analytic methods, it is also possible to find the solution through approximation techniques. Some of these methods are discussed below.

3.4.2 Spectral Methods

Spectral methods comprise of mathematical techniques used to find a solution to some types of PDEs. These methods usually express the solution of a given PDE in terms of its Fourier series which is then substituted in the PDE itself in order to obtain a system of ordinary differential equations [2, 4]. This technique often simplifies the problem. However, it is often necessary to employ numerical techniques to find the solution to each of the resulting ordinary differential equations [6].

3.4.3 Numerical Methods

The degree of difficulty presented in solving PDEs (especially those related to general physical problems) has led to the constant development of newer, faster and better numerical techniques as means for finding an approximate solution to any given PDE. The most common of these techniques are finite difference methods, finite element methods and boundary element methods [10].

- *Finite Difference Method*
    Finite difference methods are based on grid-type discretization of the unknown function in the domain in which a given PDE is solved. The derivatives involved in the PDE are then expressed in terms of these discrete points according to well established rules at every point in the grid and as many neighboring points as required by the order of the derivative in turn. It is worth mentioning that the value of the function and its derivatives are also expressed in the same manner but are somehow compensated with the boundary conditions. Then, all the corresponding expressions are substituted in the original PDE, leading to a system of algebraic equations that can be easily solved.
    Finite difference method can be further categorized into explicit, implicit and semi-implicit methods. The type of method selected to solve a particular PDE depends on the criteria often related to the inner stability of the method [12].
- *Finite Element Method*
    Finite element method is a technique that can solve either PDE or integral differential equations indistinctively. The working principle of this technique consists of approximating the original PDE by a system of ordinary differential equations that can be integrated numerically using well known methods. The main challenge when applying this technique to solve the PDE consists in approximating the original equation so that it is stable. Finite element methods are particularly useful for solving problems with a moving boundary [9].
- *Boundary Element Method*
    This method consists in finding a suitable set of boundary values by means of re-writing the original PDE as an integral equation. The boundary values found can then be used to calculate the numerical solution of the original PDE. Finite element methods are often regarded as accurate. However, they usually lead to very large matrix systems, and therefore their computational cost can be quite high [5].

3.5 Conclusions

This chapter has covered a gentle introduction to partial differential equations (PDEs). Various methods to classify the PDEs were discussed. Examples of PDEs that are commonly used in applied mathematics were also described. In particular, the use of elliptic PDEs was discussed and the Harmonic and Biharmonic equations were presented. Various solution methods available for solving PDEs, ranging from analytic to numerical schemes, were also mentioned.

References

1. Axler S, Bourdon P, Ramey W (2001) Harmonic function theory. Springer, Berlin
2. Castro CG, Ugail H, Willis P, Palmer I (2008) A survey of partial differential equations in geometric design. Vis Comput 24(3):213–225. doi:10.1007/s00371-007-0190-z
3. Evans G, Blackledge J, Yardley P (1999) Analytic methods for partial differential equations. Springer, Berlin
4. Fornberg B (1996) A practical guide to pseudospectral methods. Cambridge University Press, Cambridge
5. Gladwell I (1980) Survey of numerical methods for partial differential equations. Oxford University Press, London
6. Gottlieb D, Orzag S (1977) Numerical analysis of spectral methods: theory and applications. SIAM, Philadelphia
7. Farlow SJ (1999) Partial differential equations for scientists and engineers. Dover, New York
8. Jang CL (2011) Partial differential equations: theory, analysis and applications. Nova Publ., New York
9. Johnson C (2009) Numerical solution of partial differential equations by the finite element method. Dover, New York
10. Machura M, Sweet RA (1980) A survey of software for partial differential equations. ACM Trans Math Softw 6(4):461–488. doi:10.1145/355921.355922
11. Sapiro G (2001) Geometric partial differential equations and image analysis. Cambridge University Press, Cambridge
12. Smith GD (1985) Numerical solution of partial differential equations: finite difference methods. Clarendon, Oxford
13. Zachmanoglou EC, Thoe DW (1988) Introduction to partial differential equations with applications. Dover, New York

Chapter 4
Elliptic PDEs for Geometric Design

Abstract This chapter deals with the use of elliptic PDEs for geometric design. The chapter introduces the common elliptic PDEs such as the Laplace equation and the Biharmonic equation and shows that they can be used as a tool for surface generation. This chapter also discusses the general elliptic PDEs for surface design. Solution schemes showing how to solve the chosen elliptic PDEs in analytic form is described. Several examples of surface generation using elliptic PDEs are also given in this chapter.

4.1 Introduction

The use of elliptic PDEs for shape design is conceptually different to the conventional methods such as splines. The basic philosophy behind this method is that shape design is effectively treated as a mathematical boundary-value problem, i.e. shapes are produced by finding the solutions to a suitably chosen elliptic PDE that satisfies certain boundary conditions. Bloor and Wilson in their original papers [1, 2] illustrated how elliptic PDEs (in particular the Biharmonic equation) can be used to generate a number of smooth blending surfaces. The problem of blend generation is essentially that of generating a smooth surface which acts as a bridging surface between the neighboring primary surfaces [13]. Such a blending surface must meet the primary surface at specified curves, known as the tramlines, with a specified degree of continuity.

4.2 The Laplace Equation

Taking into consideration the discussions above, let us imagine using the standard Laplace equation to generate a blend surface. Here we consider the problem of generating a blend between a circular cylinder and a flat plane where the plane is at right angles to the cylinder. Suppose that the cylinder has radius a and that it is blended in the plane to a circular hole of radius b whereby the height of the cylinder above the plane is taken to be h. We use standard parametric coordinates $\{(u, v) : 0 \le u \le 1, 0 \le v \le 2\pi\}$. The boundary conditions for this problem are:

H. Ugail, *Partial Differential Equations for Geometric Design*,
DOI 10.1007/978-0-85729-784-6_4, © Springer-Verlag London Limited 2011

Fig. 4.1 Surface generated
as a solution to the standard
Laplace equation subject to
circular boundary conditions

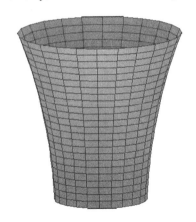

$$x(0, v) = b \cos v, \qquad y(0, v) = b \sin v, \qquad z(0, v) = 0,$$
$$x(1, v) = a \cos v, \qquad y(0, v) = a \sin v, \qquad z(0, v) = h.$$

With the above we seek a solution which is periodic in v satisfying the three equations:

$$\frac{\partial^2 x}{\partial u^2} + \frac{\partial^2 x}{\partial v^2} = 0, \tag{4.1}$$

$$\frac{\partial^2 y}{\partial u^2} + \frac{\partial^2 y}{\partial v^2} = 0, \tag{4.2}$$

$$\frac{\partial^2 z}{\partial u^2} + \frac{\partial^2 z}{\partial v^2} = 0. \tag{4.3}$$

By using the method of separation of variables the solutions for the above equations subject to the above boundary conditions are given as

$$x(u, v) = \left(b \cos u + (b - a \cosh 1) \frac{\sinh u}{\sinh 1} \right) \cos v, \tag{4.4}$$

$$y(u, v) = \left(b \cos u + (b - a \cosh 1) \frac{\sinh u}{\sinh 1} \right) \sin v, \tag{4.5}$$

$$z(u, v) = hu. \tag{4.6}$$

Taking $a = 1$, $b = 2$ and $h = 1$, Fig. 4.1 shows the resulting shape of the surface. One should note that the solution of the Laplace equation subject to the boundary conditions defines a suitable function which defines the surface blend. Since the Laplace equation is elliptic, it guarantees that the generated solution function is unique, smooth and takes its maximum and minimum values on the boundary [9].

One can clearly see that apart from the standard Laplace operator, other elliptic operators can be used. Furthermore, higher order elliptic operators can be used with, for example, derivative and curvature boundary conditions to allow higher order continuity at the boundaries.

4.2.1 Numerical Solution Using Finite Difference Method

Given the Laplace equation in parametric form defined by the two parameters u and v, the finite difference method can be employed to approximate the derivatives (of the Laplace equation) at points in the solution domain. The solution in this case is worked out by means of discretizing the (u, v) parameter space along the iso-parametric lines given by $u = u_1$ and $v = v_1$ where u_1 and v_1 are constants. One can define a set of points in the (u, v) plane as mesh points where the iso-parametric lines u_1 and v_1 intersect. The solution then involves approximating the Laplace equation at a mesh point in terms of the solution value at the mesh point itself and the solution values at neighboring points. This process forms a system of linear equations which can be solved to obtain the whole solution at all the mesh points.

Hence, given the (u, v) parameter space such that $u_0 \le u \le u_l$ and $v_0 \le v \le v_l$ we can divide it into a uniform grid of size $\delta u = h$ and $\delta u = k$ where $h = \frac{u_l - u_0}{p-1}$ and $k = \frac{v_l - v_0}{q-1}$ and $pq = n$ is the total number of mesh points.

If we denote $\mathbf{X} = (x, y, z)$ then we have $X_{ij} = (x_{ij}, y_{ij}, z_{ij})$. With this, the finite difference approximation for each coordinate space for the second order derivatives of the Laplace equation can be computed using Taylor expansion [8], e.g.

$$\left(\frac{\partial^2 x}{\partial u^2}\right)_{ij} \simeq \frac{x_{i+1,j} - 2x_{ij} + x_{i-1,j}}{h^2}, \tag{4.7}$$

$$\left(\frac{\partial^2 x}{\partial v^2}\right)_{ij} \simeq \frac{x_{i,j+1} - 2x_{ij} + x_{i,j-1}}{k^2}, \tag{4.8}$$

and the Laplace equation (for the x-component) can be approximated by the finite difference scheme as

$$\frac{x_{i+1,j} - 2x_{ij} + x_{i-1,j}}{h^2} + \frac{x_{i,j+1} - 2x_{ij} + x_{i,j-1}}{k^2} = 0. \tag{4.9}$$

4.3   The Biharmonic Equation

Similar to the Laplace equation one can also consider the Biharmonic equation,

$$\left(\frac{\partial^2}{\partial u^2} + a^2 \frac{\partial^2}{\partial v^2}\right)^2 \mathbf{X}(u, v) = 0. \tag{4.10}$$

This equation allows the conditions on $\mathbf{X}$ along with the conditions on the direction of tangents to be specified.

Again the solution of Eq. (4.10), and hence the shape of the surface, depends on both the choice of the parametric domain Ω and the parametrization of the boundary conditions. As usual the region Ω is taken to be a rectangle such that $\{\Omega : u_0 \le u \le u_1;\ v_0 \le v \le v_1\}$.

Note that the parameter a controls the relative scales of the u and v surface co-ordinates providing an additional control over the shape of the surface. For large a, changes in the u direction occur over relatively short length scale, i.e. it is $1/a$ times the length scale in the v direction over which similar changes take place [1, 12].

The boundary conditions on the solution of Eq. (4.10) relate how $\mathbf{X}(u, v)$ and its normal derivative in the (u, v) plane, $\frac{\partial \mathbf{X}}{\partial n}$, vary along $\partial \Omega$. The conditions on $\mathbf{X}$ determine the shape of the curves which bound the surface in physical space, and the derivative boundary conditions on $\frac{\partial \mathbf{X}}{\partial n}$ determine the rate and the direction in which the surface moves away from its boundaries.

4.3.1 Analytic Solution

In order to take advantage of a fast analytic solution method to generate a typical PDE surface, Eq. (4.10) is solved over a finite region Ω of the (u, v) parameter plane subject to the periodic boundary conditions on the solution $\mathbf{X}(u, v)$, which specify how $\mathbf{X}(u, v)$ and its normal derivative $\frac{\partial \mathbf{X}}{\partial \mathbf{n}}$ vary along $\partial \Omega$. With periodic boundary conditions, v being the periodic parameter, and using the method of separation of variables, the analytic solution of Eq. (4.10) can be written as

$$\mathbf{X}(u, v) = \mathbf{A}_0(u) + \sum_{n=1}^{N} \left[\mathbf{A}_n(u) \cos(nv) + \mathbf{B}_n(u) \sin(nv) \right], \qquad (4.11)$$

where

$$\mathbf{A}_0 = \mathbf{a}_{00} + \mathbf{a}_{01} u + \mathbf{a}_{02} u^2 + \mathbf{a}_{03} u^3, \qquad (4.12)$$

$$\mathbf{A}_n = \mathbf{a}_{n1} e^{anu} + \mathbf{a}_{n2} u e^{anu} + \mathbf{a}_{n3} e^{-anu} + \mathbf{a}_{n4} u e^{-anu}, \qquad (4.13)$$

$$\mathbf{B}_n = \mathbf{b}_{n1} e^{anu} + \mathbf{b}_{n2} u e^{anu} + \mathbf{b}_{n3} e^{-anu} + \mathbf{b}_{n4} u e^{-anu}, \qquad (4.14)$$

where $\mathbf{a}_{n1}, \mathbf{a}_{n2}, \mathbf{a}_{n3}, \mathbf{a}_{n4}, \mathbf{b}_{n1}, \mathbf{b}_{n2}, \mathbf{b}_{n3}$ and $\mathbf{b}_{n4}$ are vector-valued constants, whose values are determined by the imposed boundary conditions at $u = u_0$ and $u = u_1$. Note that N is a positive integer value.

In the case of the Biharmonic equation, to determine its solution (4.11), and hence to create a surface, it is necessary to define a set of four boundary conditions: two positional boundary conditions and two derivative boundary conditions. Taking Ω to be the region such that $\{\Omega : 0 \le u \le 1; \ 0 \le v \le 2\pi\}$, Eq. (4.10) is solved with the boundary conditions imposed on the solution of the form:

$$\mathbf{X}(0, v) = \mathbf{p}_1(v), \qquad (4.15)$$

$$\mathbf{X}(1, v) = \mathbf{p}_2(v), \qquad (4.16)$$

$$\mathbf{X}_u(0, v) = \mathbf{d}_1(v), \qquad (4.17)$$

$$\mathbf{X}_u(1, v) = \mathbf{d}_2(v). \qquad (4.18)$$

The boundary conditions $\mathbf{p}_1(v)$ and $\mathbf{p}_2(v)$ define the edges of the surface patch at $u = 0$ and $u = 1$, respectively, as parameterized in terms of the v variable. These conditions are termed the positional boundary conditions. The conditions $\mathbf{d}_1(v)$ and $\mathbf{d}_2(v)$, termed derivative boundary conditions, can be used to determine the surface

Fig. 4.2 The effect of derivative conditions on a Biharmonic PDE surface

normals at the corresponding boundaries of the surface. The derivative conditions play an important role in determining the overall shape of the surface.

For example, Fig. 4.2 shows a sequence of surfaces which illustrate the effect of the derivative condition for $\mathbf{X}_u$ on the shape of the surface. Note that all the surfaces shown in Fig. 4.2 have the same boundary conditions on the function $\mathbf{X}$ whereas the boundary condition on the function $\mathbf{X}_u$ at $u = 1$ (the top boundary in the figure), in particular, its direction has been varied (by means of considering circular functions of varying radii). This illustrates the fact that the derivative conditions control the direction along which the surface leaves the boundary curves.

One should note that, in order to utilize the solution given in (4.11), the boundary conditions must be given in terms of a periodic function which is equivalent to a Fourier series with a given number of Fourier modes N. This is indeed a restriction on the above analytic solution; however, a surprisingly wide variety of geometric shapes can be generated using this solution scheme whereby the boundary conditions are given as a Fourier series.

In some cases, the boundary conditions are defined as curves in $\mathbf{R}^3$, which often can be in terms of a discrete set of points. For example, it is often the case that the boundary conditions for a surface patch are defined in terms of a cubic B-spline of the form

$$S(v) = \sum_i \underline{c}_i B_i(v). \tag{4.19}$$

In such cases, in order to utilize the solution given in (4.11), it is necessary to approximate the curves by discretely sampling them at regular intervals and performing discrete Fourier analysis. This process is briefly described below.

4.3.1.1 Discrete Fourier Analysis

Given a function $f(v)$ in the form

$$f(v) = a_0 + \sum_{n=1}^{\infty} \left[a_n \cos(nv) + b_n \sin(nv) \right], \tag{4.20}$$

the coefficients a_0, a_n and b_n can be defined as

$$a_0 = \frac{2}{\pi} \int_0^{2\pi} f(v)\,dv, \tag{4.21}$$

$$a_n = \frac{1}{\pi} \int_0^{2\pi} f(v)\cos(nv)\,dv, \tag{4.22}$$

$$b_n = \frac{1}{\pi} \int_0^{2\pi} f(v)\sin(nv)\,dv. \tag{4.23}$$

Thus, given a periodic function $g(v)$, in order to approximate the function in the form given in Eq. (4.20), one would need to approximate the integrals given in Eqs. (4.21), (4.22) and (4.23).

There are several methods to approximate the integrals and each of them is equivalent to finding the area under the curve. If we sample the periodic function $g(v)$ at n points of regular intervals of the angle s then

$$\int_0^{2\pi} g(v)\,dv \simeq s(g_0 + g_1 + g_2 + \cdots + g_{n-1}). \tag{4.24}$$

The accuracy of the computation depends on the sampling number.

Given $g(v)$ as a set of evenly spaced function values, a simpler way of computing the integrals is by finding the mean value of the functions $g(v)$, $g(v)\cos(nv)$ and $g(v)\sin(nv)$ for successive n, over the complete cycle of the sample space [5], i.e.

$$\frac{2}{\pi} \int_0^{2\pi} g(v)\,dv = 2 \times \text{mean value of } g(v) \text{ over a period,} \tag{4.25}$$

$$\frac{1}{\pi} \int_0^{2\pi} g(v)\cos(nv)\,dv = 2 \times \text{mean value of } g(v)\cos nv \text{ over a period,} \tag{4.26}$$

$$\frac{1}{\pi} \int_0^{2\pi} g(v)\sin(nv)\,dv = 2 \times \text{mean value of } g(v)\sin nv \text{ over a period.} \tag{4.27}$$

4.3.2 Geometric Properties of the Biharmonic PDE

The Biharmonic operator can be seen to act as a smoothing operator which enables producing an interpolating surface for a given set of boundary data. The resulting surface in the above case is provided as an analytic expression and is infinitely differentiable. An important point to highlight here is that, since we are treating surface generation as a boundary-value problem, the resulting surface is entirely dependent on the boundary conditions, and hence the boundary conditions can be utilized as a surface manipulation tool.

Common parametric surface generation methods such as those based on spline techniques have attractive geometric properties through which the behavior of the surface subject to changes in the relating control points are somewhat intuitive. For example, in the case of Bézier surfaces, the convex hull property guarantees that the resulting surface is entirely bounded within the convex hull of the control polygons which determine the shape of the surface. In the case of surfaces generated as solutions of PDEs, in particular low order elliptic PDEs, similar geometric properties can be identified. Thus, in the case of PDE surfaces discussed here, one can also show that the resulting surface behaves in an intuitive fashion subject to changes in the boundary data.

As discussed in the previous section for the Laplace equation, one can show (through the min/max principle) that the maximum/minimum of the interpolating function occurs at the boundaries of the surface patch [9]. Whilst this desirable property holds for the Laplace equation, the surfaces generated by it are somewhat limited since there are only two boundary conditions available for the user to define a surface patch. The Biharmonic equation, on the other hand, is more powerful since the user can impose four boundary conditions, two defining the edges of the surface patch and the two defining the rate of change of these edges which determine the interior of the surface patch. Though there is no maximum/minimum principle for the Biharmonic equation, one can still find a priori estimates on the bounds of the interpolating function resulting from the Biharmonic equation. This can be undertaken by applying the maximum/minimum principle for the quantity $\| \nabla \mathbf{X} \|^2 - \mathbf{X} \nabla^2 \mathbf{X}$ and applying the maximum modulus theorem [10] so that

$$\| \mathbf{X} \| \le K \left(\| \mathbf{X}_0 \| + \| (\partial \mathbf{X}/\partial n)_0 \| \right), \tag{4.28}$$

where Eq. (4.10) is solved over the region Ω subject to the conditions $\mathbf{X} = \mathbf{X}_0$ and $(\partial \mathbf{X}/\partial n) = (\partial \mathbf{X}/\partial n)_0$ on the boundary of Ω. Here K is a constant which depends only on the Biharmonic operator and the geometric shape of Ω. Equation (4.28) indicates that the resulting surface will be of order corresponding to the maximum dimensions of the boundary conditions summed with the maximum rate of change of distance with which the surface moves away from the boundary.

4.4   General Elliptic PDEs

The boundary-value approach on which the elliptic PDE model is based can be described as follows. Assuming we employ the usual parametric coordinates such that $\mathbf{X}(u, v)$ is a parametric function which gives the surface in $\mathbf{R}^3$, we can regard $\mathbf{X}$ as the solution of a PDE of the form

$$D_{u,v}^m(\mathbf{X}) = \mathbf{F}(u, v), \tag{4.29}$$

where $D_{u,v}^m$ is a partial differential operator of order m in independent variables u and v, and $\mathbf{F}$ is a vector-valued function of u and v.

Similar to the discussion regarding the Biharmonic operator, the general partial differential operator in Eq. (4.29) is a smoothing operator in which the function values at any point on the surface represents, in a sense, a weighted average of the surrounding values. Thus, a surface is obtained as an average of the imposed boundary conditions. Furthermore, surfaces generated using this technique are fair [11] in the sense that they do not possess small scale oscillations, provided, of course, such oscillations do not exist in the boundary conditions. Also, excluding the possibility of discontinuities at the boundary conditions, PDE surfaces are infinitely differentiable (even for discontinuities on boundaries, the function $\mathbf{X}(u, v)$ is differentiable) on the surface interior.

In order to generalize the formulation of the Laplace and Biharmonic equation to a general PDE of elliptic type, we proceed as follows. For the general case, we require an elliptic PDE that satisfies a given number of $2N$ boundary conditions. Here N is an arbitrary integer such that $N \geq 2$. The boundary conditions can be given in the form,

$$\mathbf{X}(0, v) = \mathbf{f}_1(v), \tag{4.30}$$

$$\mathbf{X}(u_i, v) = \mathbf{g}_i(v), \quad i = 2, \dots, 2N - 1, \tag{4.31}$$

$$\mathbf{X}(1, v) = \mathbf{f}_{2N}(v), \tag{4.32}$$

where $\mathbf{f}_1(v)$ in Eq. (4.30) and $\mathbf{f}_{2N}(v)$ in Eq. (4.32) are function boundary conditions specified at $u = 0$ and $u = 1$, respectively. The conditions $\mathbf{X}(u_i, v) = \mathbf{g}_i(v)$ in Eq. (4.31) can take the form of either

$$\mathbf{X}(u_i, v) = \mathbf{f}_i(v) \quad \text{for } 0 < u_i < 1, \ i = 2, \dots, 2N - 1, \tag{4.33}$$

or may involve

$$\frac{\partial \mathbf{X}}{\partial u}, \frac{\partial^2 \mathbf{X}}{\partial u^2}, \frac{\partial^3 \mathbf{X}}{\partial u^3}, \dots, \frac{\partial^{2N-2} \mathbf{X}}{\partial u^{2N-2}} \quad \text{for } 0 \leq u_i \leq 1, \ i = 2, \dots, 2N - 1. \tag{4.34}$$

In simpler terms, the above boundary condition implies that for a PDE surface patch of order $2N$ we can specify two function boundary conditions, as given in Eqs. (4.30) and (4.32), that should be satisfied at the edges (at $u = 0$ and $u = 1$) of the surface patch, and a number of function or derivative conditions, as given in Eq. (4.31), amounting to $2N - 2$ boundary conditions which the PDE should also satisfy.

With the above formulation, we take the standard Laplace equation, $\nabla^2 \mathbf{X} = 0$, as a base PDE and generalize it to the Nth order such that

$$\left(\frac{\partial^2}{\partial u^2} + a^2 \frac{\partial^2}{\partial v^2} \right)^N \mathbf{X}(u, v) = 0. \tag{4.35}$$

4.4.1 Analytic Solution

Given a set of $2N$ boundary conditions as defined in Eqs. (4.30), (4.31) and (4.32), we take the (u, v) parameter space Ω to be the region $\{u, v : 0 \leq u \leq 1; 0 \leq v \leq 2\pi\}$. Thus, we assume that all the boundary conditions are periodic in v in the sense $\mathbf{f}_0(0) = \mathbf{f}_0(2\pi)$, $\mathbf{f}_1(0) = \mathbf{f}_1(2\pi)$ and $\mathbf{g}_i(0) = \mathbf{g}_i(2\pi)$. We further assume that all the boundary conditions are continuous functions within the domain Ω.

With the above assumptions on the boundary conditions, we can utilize the method of separation of variables to write the analytic solution of Eq. (4.35) as

$$\mathbf{X}(u, v) = \mathbf{A}_0(u) + \sum_{n=1}^{\infty} \left[\mathbf{A}_n(u) \cos(nv) + \mathbf{B}_n(u) \sin(nv) \right], \qquad (4.36)$$

where

$$\mathbf{A}_0 = \mathbf{a}_{00} + \mathbf{a}_{01}u + \mathbf{a}_{02}u^2 + \cdots + \mathbf{a}_{(2N-1)}u^{2N-1}, \qquad (4.37)$$

$$\mathbf{A}_n = \mathbf{a}_{n1}e^{anu} + \mathbf{a}_{n2}ue^{anu} + \mathbf{a}_{n3}e^{-anu} + \mathbf{a}_{n4}ue^{-anu} + \cdots$$
$$+ \mathbf{a}_{n(2N-3)}u^{N-2}e^{anu} + \mathbf{a}_{n(2N-2)}u^{N-1}e^{anu} + \mathbf{a}_{n(2N-1)}u^{N-2}e^{-anu}$$
$$+ \mathbf{a}_{n2N}u^{N-1}e^{-anu}, \qquad (4.38)$$

$$\mathbf{B}_n = \mathbf{b}_{n1}e^{anu} + \mathbf{b}_{n2}ue^{anu} + \mathbf{b}_{n3}e^{-anu} + \mathbf{b}_{n4}ue^{-anu} + \cdots$$
$$+ \mathbf{b}_{n(2N-3)}u^{N-2}e^{anu} + \mathbf{b}_{n(2N-2)}u^{N-1}e^{anu} + \mathbf{b}_{n(2N-1)}u^{N-2}e^{-anu}$$
$$+ \mathbf{b}_{n2N}u^{N-1}e^{-anu}, \qquad (4.39)$$

where $\mathbf{a}_{00} + \mathbf{a}_{01}$, $\mathbf{a}_{02}$, ..., $\mathbf{a}_{2N-1}$, $\mathbf{a}_{n1} + \mathbf{a}_{n2}$, $\mathbf{a}_{n3}$, $\mathbf{a}_{n4}$, ..., $\mathbf{a}_{n(2N-3)}$, $\mathbf{a}_{n(2N-2)}$, $\mathbf{a}_{n(2N-1)}$, $\mathbf{a}_{n2N}$ and $\mathbf{b}_{n1} + \mathbf{b}_{n2}$, $\mathbf{b}_{n3}$, $\mathbf{b}_{n4}$, ..., $\mathbf{b}_{n(2N-3)}$, $\mathbf{b}_{n(2N-2)}$, $\mathbf{b}_{n(2N-1)}$, $\mathbf{b}_{n2N}$ are vector-valued constants, whose values are determined by the imposed boundary conditions at u_i where $0 \le u_i \le 1$.

Since the chosen boundary conditions are all continuous functions which are also periodic in v, we can write down their Fourier series representation as

$$\mathbf{f}_1(v) = \mathbf{C}_0^1 + \sum_{n=1}^{\infty} \left[\mathbf{C}_n^1 \cos(nv) + \mathbf{S}_n^1 \sin(nv) \right], \qquad (4.40)$$

$$\mathbf{g}_i(v) = \mathbf{C}_0^i + \sum_{n=1}^{\infty} \left[\mathbf{C}_n^i \cos(nv) + \mathbf{S}_n^i \sin(nv) \right], \quad i = 2, \ldots, 2N - 1, \quad (4.41)$$

$$\mathbf{f}_{2N}(v) = \mathbf{C}_0^{2N} + \sum_{n=1}^{\infty} \left[\mathbf{C}_n^{2N} \cos(nv) + \mathbf{S}_n^{2N} \sin(nv) \right]. \qquad (4.42)$$

If we now assume for the moment that the Fourier sums in the Expressions (4.40), (4.41) and (4.42) have finite M modes, then the vector constants $\mathbf{C}_0^1$, $\mathbf{C}_0^i$ for $i = 2, \ldots, 2N - 1$, and $\mathbf{C}_0^{2N}$ can be obtained by directly comparing them with the constants $\mathbf{a}_{00}$, $\mathbf{a}_{01}$, $\mathbf{a}_{02}$, ..., $\mathbf{a}_{(2N-1)}$ given in Eq. (4.37). Now for each of the Fourier modes $n \in \{1, \ldots, M\}$ we can write linear systems

$$\begin{pmatrix} \mathbf{a}_{n1} \\ \vdots \\ \mathbf{a}_{2N} \end{pmatrix} = \mathbf{A}(a, n) \begin{pmatrix} \mathbf{C}_n^1 \\ \vdots \\ \mathbf{C}_n^{2N} \end{pmatrix} \qquad (4.43)$$

and

$$\begin{pmatrix} \mathbf{b}_{n1} \\ \vdots \\ \mathbf{b}_{2N} \end{pmatrix} = \mathbf{B}(a, n) \begin{pmatrix} \mathbf{S}_n^1 \\ \vdots \\ \mathbf{S}_n^{2N} \end{pmatrix}, \qquad (4.44)$$

where $\mathbf{A}(a, n)$ and $\mathbf{B}(a, n)$ are $2N \times 2N$ matrices whose coefficients can be obtained by solving the linear systems (4.43) and (4.44), subject to the Fourier coefficients corresponding to the $2N$ boundary conditions.

The above solution scheme is based on the fact that the boundary conditions can be expressed as finite Fourier series. However, in practical design scenarios such an assumption is not a realistic proposition. In order to take care of this, one can adopt a generalized version of the spectral approximation to the Biharmonic PDE given in [3, 4] as described below.

Although in practical terms we cannot assume that a given boundary condition can be expressed accurately using a finite Fourier series, it is reasonable to assume that the boundary conditions can be written as,

$$\mathbf{f}_1(v) = \mathbf{C}_0^1 + \sum_{n=1}^{M} \left[\mathbf{C}_n^1 \cos(nv) + \mathbf{S}_n^1 \sin(nv) \right] + \mathbf{R}_1(v), \tag{4.45}$$

$$\mathbf{g}_i(v) = \mathbf{C}_0^i + \sum_{n=1}^{M} \left[\mathbf{C}_n^i \cos(nv) + \mathbf{S}_n^i \sin(nv) \right] + \mathbf{R}_i(v), \quad i = 2, \dots, 2N - 1, \tag{4.46}$$

$$\mathbf{f}_{2N}(v) = \mathbf{C}_0^{2N} + \sum_{n=1}^{M} \left[\mathbf{C}_n^{2N} \cos(nv) + \mathbf{S}_n^{2N} \sin(nv) \right] + \mathbf{R}_{2N}(v). \tag{4.47}$$

Thus, the basic idea here is to formulate each of the boundary conditions in terms of the sum of a finite Fourier series containing M modes and a 'remainder' term $\mathbf{R}_i(v)$, $i = 1, \dots, 2N$, which contains the higher order Fourier modes. For the case of Biharmonic equation, it is shown that the higher order Fourier modes make negligible contributions to the interior of the PDE patch, and the same applies for the general case of the Nth order Biharmonic PDE [3]. Hence it is reasonable to truncate the Fourier series at some finite M (typically $M = 6$ is adequate) and represent the contribution of the high frequency modes to the surface with a remainder function $\mathbf{R}(u, v)$. The format of this remainder function is somewhat arbitrary and for this work it is taken to be of the form

$$\begin{aligned} \mathbf{R}(u, v) = {} & \mathbf{r}_1 e^{\omega u} + \mathbf{r}_2 u e^{\omega u} + \mathbf{r}_3 e^{-\omega u} + \mathbf{r}_4 u e^{-\omega u} + \cdots \\ & + \mathbf{r}_{(2N-3)} u^{N-2} e^{\omega u} + \mathbf{r}_{(2N-2)} u^{N-1} e^{\omega u} + \mathbf{r}_{(2N-1)} u^{N-2} e^{-\omega u} \\ & + \mathbf{r}_{2N} u^{N-1} e^{-\omega u}, \end{aligned} \tag{4.48}$$

where $\mathbf{r}_1, \mathbf{r}_2, \dots, \mathbf{r}_{(2N-1)}, \mathbf{r}_{2N}$ are vector-valued constants which depend on v.

Now by taking $\widetilde{\mathbf{X}}(u, v)$ to be the approximate solution,

$$\widetilde{\mathbf{X}}(u, v) = \mathbf{A}_0(u) + \sum_{n=1}^{M} \left[\mathbf{A}_n(u) \cos(nv) + \mathbf{B}_n(u) \sin(nv) \right], \tag{4.49}$$

satisfying the boundary conditions of the finite Fourier series, we define difference functions such that

$$\mathbf{df}_1(v) = \mathbf{df}_1(v) - \widetilde{\mathbf{X}}(0, v), \tag{4.50}$$

$$\mathbf{dg}_i(v) = \mathbf{g}_i(v) - \widetilde{\mathbf{X}}(u_i, v), \quad i = 2, \ldots, 2N-1, \tag{4.51}$$

$$\mathbf{df}_{2N}(v) = \mathbf{df}_1(v) - \widetilde{\mathbf{X}}(1, v). \tag{4.52}$$

By choosing ω in Expression (4.48) to be an, the vector constants $\mathbf{r}_1, \ldots, \mathbf{r}_{2N}$ can be computed by means of direct comparison with the difference terms $\mathbf{df}_1(v)$, $\mathbf{df}_{2N}(v)$ and $\mathbf{dg}_i(v)$, for $i = 2, \ldots, 2N-1$ in Eqs. (4.50)–(4.52).

Finally, the approximate analytic solution of the PDE is given as

$$\mathbf{X}(u, v) = \widetilde{\mathbf{X}}(u, v) + \mathbf{R}(u, v). \tag{4.53}$$

It is important to note that the choice of the number of Fourier terms M will affect how well $\widetilde{\mathbf{X}}(u, v)$ approximates the solutions of the general Nth order Biharmonic PDE. Although this may be the case, due to the choice of difference functions utilized here, i.e. by computing the difference between the original boundary conditions and the corresponding finite Fourier series, as described in Eqs. (4.50)–(4.52), the approximate solution satisfies the chosen set of boundary conditions exactly, to within the machine accuracy.

4.5   Other Variations of the General Elliptic Equation

The original formulation of the PDE method makes use of a fourth order elliptic PDE. However, alternatives for such a formulation have been developed throughout [6, 7, 15]. Such alternatives have been developed in order to fulfill specific requirements relating to a particular application. For instance, sixth order elliptic PDEs have been used in order to guarantee continuity in the curvature throughout the PDE surface. This formulation uses the following elliptic PDE

$$\left(a \frac{\partial^6}{\partial u^6} + b \frac{\partial^6}{\partial u^4 \partial v^2} + c \frac{\partial^6}{\partial u^2 \partial v^4} + d \frac{\partial^6}{\partial v^6} \right) \mathbf{X}(u, v) = 0, \tag{4.54}$$

where a, b, c and d are shape control parameters. The reader can see that the differences include the order of the elliptic PDE, which in this case is of the sixth order, and the four shape control parameters within the PDE itself instead of just one as established in the original formulation [14, 15].

Another modification carried out to the standard formulation of the PDE method is based on the use of the four boundary conditions determined by the user as positional boundary conditions. That is, taking for an example the Biharmonic equation, the same positional boundary conditions $\mathbf{p}_1$ and $\mathbf{p}_2$ are used as before, but the derivative conditions $\mathbf{d}_1$ and $\mathbf{d}_2$ are also used as positional boundary conditions at $u = u_2$ and $u = u_3$, respectively, with $0 \le u_2 \le u_3 \le 1.0$. This requires a number of modifications to the matrices defining the systems of algebraic equations to be solved. However, the core of the methodology followed to find the solution of the PDE remains unchanged. These modifications can be carried out to satisfy special needs of particular applications in which the user is more interested in obtaining a surface passing through all the given curves.

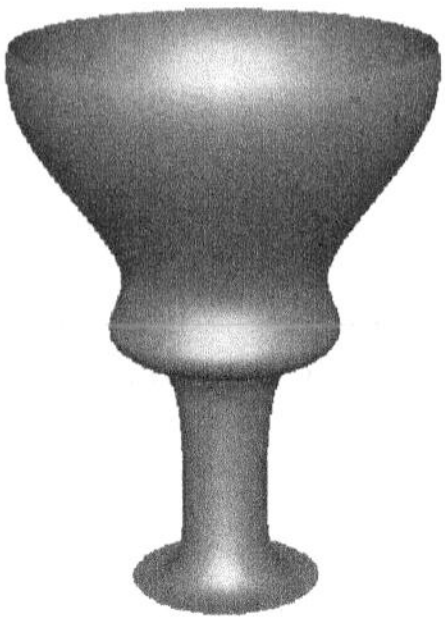

Fig. 4.3 Geometry of a
cup-like shape using the
Biharmonic equation

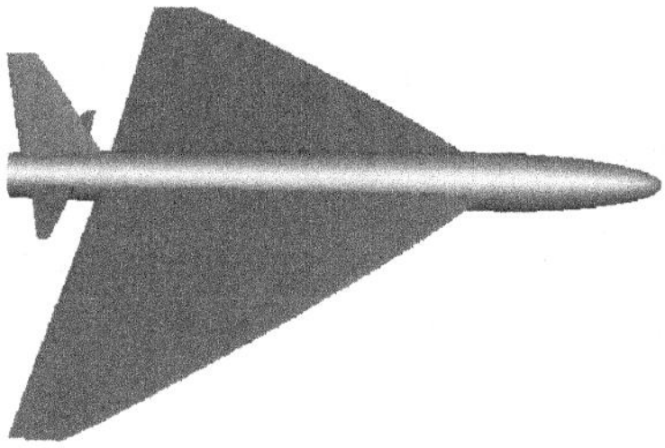

Fig. 4.4 Shape of a delta
airplane surface generated
using eight surface patches
with common boundary
conditions using the
Biharmonic equation

4.6 Examples

In this section, we discuss a number of examples of generating complex geometry
using elliptic PDEs discussed above.

Figure 4.3 shows the geometry of a cup-like shape. This shape is generated using
two PDE surface patches with a common boundary condition. The boundary conditions are defined in terms of finite Fourier series which are utilized to generate
the function definition of the surfaces through the Biharmonic equation. Here the
surface patches meet each other at the common boundaries, the derivative boundary
conditions were chosen to ensure that the surfaces are blended together with tangent
continuity.

Figure 4.4 shows the shape of a B17 airplane generated using eight surface
patches with common boundary conditions. The shape generating PDE utilized here
is again the Biharmonic equation where the boundary conditions are defined in terms
of finite Fourier series. The fuselage shape in this case is generated using a single
surface patch to which the wing surface patches are blended. In a similar fashion,
surface patches corresponding to the tail part of the airplane are generated. Additionally, two separate surfaces are then generated corresponding to the two engine
shapes.

Figure 4.5 shows the shape of Klein bottle generated as using the general elliptic
PDE. As for the boundary conditions, here we have taken cross-sectional curves of
the Klein bottle defined in analytic form given by

$$x = \begin{cases} \alpha \cos(u)(1 + \sin(u)) + \gamma \cos(u)\cos(v) & \text{if } 0 \leq u < \pi, \\ \alpha \cos(u)(1 + \sin(u)) + \gamma \cos(v + \pi) & \text{if } \pi \leq u \leq 2\pi, \end{cases} \quad (4.55)$$

Fig. 4.5 Shape of a Klein
bottle generated using the
general elliptic PDE or order
40

Fig. 4.6 Original scan data
corresponding to a 3D face

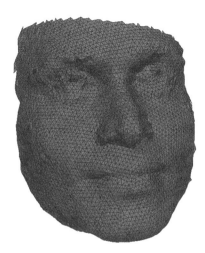

$$y = \begin{cases} \beta \sin(u) + \gamma \sin(u) \cos(v) & \text{if } 0 \le u < \pi, \\ \beta \sin(u) & \text{if } \pi \le u \le 2\pi, \end{cases} \quad (4.56)$$

$$z = \gamma \sin(v) \quad (4.57)$$

where 40 cross-sectional ellipses along u were taken with $0 \le v \le 2\pi$, $\alpha = 6$, $\beta = 16$ and $\gamma = 4(1 - \cos(u)/2)$. In order to generate the full Klein bottle shape, the 40th order general elliptic PDE was solved.

As a last example in this section, we discuss how an existing geometry data can be represented using PDEs. Here we take the case of representing a face given the corresponding data from a 3D laser scanner.

Figure 4.6 shows a typical facial data set which is obtained from a 3D scanner. The raw data from the 3D scanner has the information on the location of data points in the physical space and a triangular connectivity defined between these points. Taking this data, one can define a series of planes through which the data passes. For each such plane, one can then identify the points from scan data which belong to the plane. Hence, using this technique, a series of profile curves can be automatically extracted.

Fig. 4.7 Data points
corresponding to an ordered
set of curves for defining the
PDE boundary conditions

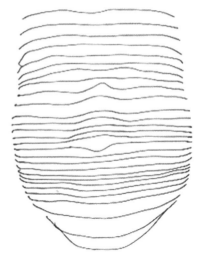

Fig. 4.8 PDE based
reconstruction of the face

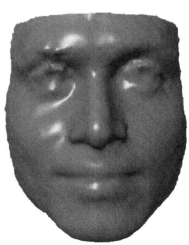

Figure 4.7 shows such profile curves which have been automatically extracted for data corresponding to the face shown in Fig. 4.5. Once the curves are extracted, for each curve we then fit a cubic spline. This process of curve fitting to the extracted discrete curve data enables having a smooth curve passing through the discrete data as well having an equal number of curve points for each profile curve. Data points at regular intervals from the spline curves are then sampled to obtain the necessary Fourier series which are taken as boundary conditions for the PDE.

Once the discrete Fourier series for the profile curves are available, they are then categorized into groups of four and the Biharmonic equations with function boundary conditions are solved appropriately to generate the shape of the face as shown in Fig. 4.8.

4.7 Conclusions

This chapter has introduced the idea of boundary-value approach to surface design whereby elliptic PDEs are used subject to a certain set of boundary conditions. Common elliptic PDEs and their generalizations have been considered. Solution schemes, particulary Fourier based fast analytic solution schemes have been discussed.

References

1. Bloor MIG, Wilson MJ (1989) Generating blend surfaces using partial differential equations. Comput Aided Des 21(3):33–39. doi:10.1016/0010-4485(89)90071-7
2. Bloor MIG, Wilson MJ (1989) Blend design as a boundary-value problem. In: Straßer W, Seidel HP (eds) Theory and practise of geometric modelling. Springer, Berlin, pp 221–234
3. Bloor MIG, Wilson MJ (1996) Spectral approximations to PDE surfaces. Comput Aided Des 82(2):145–152. doi:10.1016/0010-4485(95)00060-7
4. Bloor MIG, Wilson MJ (2005) An analytic pseudo-spectral method to generate a regular 4-sided PDE surface patch. Comput Aided Geom Des 22(3):203–219. doi:10.1016/j.cagd.2004.08.005
5. Brigham EO (1988) The fast Fourier transform and its applications. Prentice Hall, New York
6. Du H, Qin H (2000) Direct manipulation and interactive sculpting of PDE surfaces. Comput Graph Forum 19(3):261–270. doi:10.1111/1467-8659.00418
7. Du H, Qin H (2007) Free-form geometric modeling by integrating parametric and implicit PDEs. IEEE Trans Vis Comput Graph 13(3):549–561. doi:10.1109/TVCG.2007.1004
8. Greenberg M (1998) Advanced engineering mathematics. Prentice Hall, New York
9. Reed M, Simon B (1978) Methods of modern mathematical physics IV: analysis of operators. Academic Press, San Diego
10. Sperb R (1981) Maximum principles and their applications. Academic Press, San Diego
11. Taubin G (1995) A signal processing approach to fair surface design. In: Computer Graphics Proceedings, pp 351–358. doi:10.1145/218380.218473
12. Ugail H, Bloor MIG, Wilson MJ (1999) Techniques for interactive design using the PDE method. ACM Trans Graph 18(2):195–212. doi:10.1145/318009.318078
13. Woodward CD (1987) Blends in geometric modelling. In: Martin RR (ed) Mathematical methods of surfaces II. Oxford University Press, London, pp 255–297
14. Zhang JJ, You LH (2001) Surface representation using second fourth and mixed order partial differential equations. In: International conference on shape modeling and applications, Genova, Italy
15. Zhang JJ, You LH (2002) PDE based surface representation—vase design. Comput Graph 26(1):89–98. doi:10.1016/S0097-8493(01)00160-1

Chapter 5
Interactive Design

Abstract Using elliptic PDEs described in the previous chapter, especially using the Biharmonic equation, one can create the shape of an initial surface. This can be carried out through the interactive specification of curves which can be taken as the boundary conditions for the chosen PDE. Once this is done, it may be necessary to further manipulate the geometry in order to improve the shape. Hence, it is desirable to have as much control as possible over the shape of the surface once it has been defined.

5.1 The Approach to Interactive Surface Design

In this section, we discuss how one can develop a system based on PDEs to create a physically realistic object. It is noteworthy, like many other geometric design software users, the person who may use a geometric design system based on PDEs may not necessarily know how to solve complex PDEs, nor would she/he necessarily be familiar with the influence of the boundary conditions on the solution of the chosen PDE. Therefore, a methodology must exist which allows the user to operate in an instinctive way to produce the desired surface shape. In other words, the details of the mathematical resolution of the underlying boundary-value problem must be hidden from view.

If we take the standard Biharmonic equation as an example of an interactive design tool, then, in order to determine the solution of Eq. (4.10), and hence to create a surface, it is necessary to define a set of four boundary conditions, i.e. two positional boundary conditions and two derivative boundary conditions. The positional boundary conditions determine the shape of the boundaries or the edges of the surface patch, and the derivative boundary condition determine the direction along which the surface propagates from the edges into the interior of the surface. This can be done by means of defining the boundary conditions in terms of curves in $\mathbf{R}^3$. Thus, to enhance the ease with which the user can specify the boundary conditions, and for the user to be able to readily appreciate the influence of the boundary conditions on the shape of the surface, we need to define the boundary conditions in terms of curves in $\mathbf{R}^3$.

Here one can implement a technique to create a boundary curve in $\mathbf{R}^3$, within a graphical user interface where the user is able to interact with the boundary con-

H. Ugail, *Partial Differential Equations for Geometric Design*,
DOI 10.1007/978-0-85729-784-6_5, © Springer-Verlag London Limited 2011

Fig. 5.1 An example
boundary curve which can be
used to define the PDE
boundary conditions
interactively

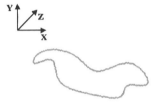

ditions and other associated parameters using the usual computer hardware devices
such as the mouse.

One method of creating curves in $\mathbf{R}^3$ is using B-splines. Consider the space curve
lying in the xy plane which is defined by means of a cubic B-spline of the form

$$S(v) = \sum_i \mathbf{c}_i B_i(v),   \tag{5.1}$$

where the control points $\mathbf{c}_i$ can be used to manipulate the curve, i.e., to change the
shape of this curve in the xy plane, the control points can be moved around with the
aid of the mouse. Once a desired shape of the curve in the xy plane is achieved, a
different plane (xz or yz) can be chosen where the curve can be further manipulated,
by moving around the control points, without changing the projection of the curve in
the original xy plane. An example of a curve that results from this process is shown
in Fig. 5.1.

Apart from defining the boundary curves using B-spline functions, a library of
standard plane curves, such as the circle and the ellipse, can also be created, which
the user would able to select by clicking on a menu using the mouse. Various geometric transformations such as translations and rotations can then be applied to such
a curve in order to create a curve in $\mathbf{R}^3$.

Figure 5.2 shows examples of curves in $\mathbf{R}^3$ created interactively. Such curves
can be used to define the necessary boundary conditions of the PDE as discussed
below. The curve at $u = 0$ corresponds to $\mathbf{p}_1(v)$, and that at $u = 1$ corresponds to
$\mathbf{p}_2(v)$. For convenience, the curves are marked p_1 and p_2 where p_1 corresponds to
the boundary condition $\mathbf{p}_1(v)$ and p_2 corresponds to the boundary condition $\mathbf{p}_2(v)$.

Once the desired shape of the curve (which represents a 'character-line' of the
desired shape of the surface) is created, we can obtain an ordered set of the points in
$\mathbf{R}^3$ on the curve. However, such a discrete representation of the curve (i.e. in terms of
coordinates of the points) is not in a suitable form with which to define the boundary
conditions to solve the PDE by the solution method described earlier. In particular,
the coefficients of the Fourier modes representing the curve are what is required.
To obtain these coefficients, a discrete Fourier transform of the curve needs to be
performed, and for this purpose it is important to choose a suitable parametrization
for the curve. One way to do this is to parameterize the curves in terms of arc length
in terms of the variable v.

Referring to the curves in Fig. 5.2(a), the point marked by a cross on each curve is
the position of the point where $v = 0$. This point serves as a rather useful parameter,
which controls v-parametrization [4, 9].

Fig. 5.2 Illustration of how
to choose the PDE boundary
conditions for the Biharmonic
equation

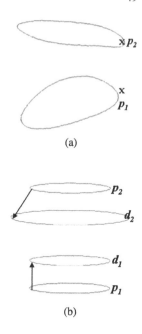

(a)

(b)

Using the above described procedures, a user is able to define a space curve and
position it in $\mathbf{R}^3$. Hence a user is effectively able to define the boundary conditions
on the function $\mathbf{X}(u, v)$.

Described next is a technique which can be implemented in order to define the
boundary conditions $\mathbf{d}_1(v)$ and $\mathbf{d}_2(v)$ on the function $\frac{\partial \mathbf{X}}{\partial u}$, which is the normal
derivative of $\mathbf{X}$ on the boundaries at $u = 0$ and $u = 1$, respectively. There are several
ways by which the boundary conditions on $\mathbf{X}_u$ can be defined. One way to do this
is to define a vector field, which is referred to as the 'derivative vector', along the
positional boundary curves. To define such a derivative vector, one can create a new
curve in $\mathbf{R}^3$ near each boundary curve at $u = 0, 1$. The difference between each point
on this newly defined 'derivative' curve (where there is a one-to-one association of
points on the two curves) and an associated point on the curve corresponding to the
positional boundary conditions determines both the magnitude (to within a scaling
factor s) and the direction of the derivative vector. In other words, the vector fields,
corresponding to the difference between the points on the curves marked $\mathbf{p}_1$ and $\mathbf{p}_2$
and those marked $\mathbf{d}_1$ and $\mathbf{d}_2$, respectively, define the function $\frac{\partial \mathbf{X}}{\partial u}$. Thus,

$$\frac{\partial \mathbf{X}}{\partial u} = \big[\mathbf{p}(v) - \mathbf{d}(v)\big]s, \tag{5.2}$$

which serves as the derivative boundary conditions. Figure 5.2 illustrates how the
derivative vector is defined by this method. For convenience, the derivative curves
are marked $\mathbf{d}_1$ and $\mathbf{d}_2$, where $\mathbf{d}_1$ corresponds to the boundary condition $\mathbf{d}_1(v)$ and
$\mathbf{d}_2$ corresponds to the boundary condition $\mathbf{d}_2(v)$.

As previously mentioned, once the boundary (both positional and derivative)
curves are defined, discrete Fourier transforms of the curves are then performed to

Fig. 5.3 PDE boundary
curves and the corresponding
surface

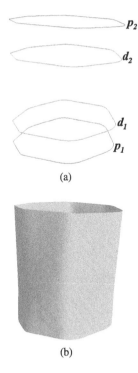

(a)

(b)

determine the coefficients of the Fourier modes. These coefficients, which determine
the necessary boundary conditions of the PDE, are used to compute the solution of
the PDE. From the solution of the PDE, a rectangular mesh of points can be obtained, which can be used to render or represent the surface. Since the (u, v) mesh
in $\mathbf{R}^2$ is an ordered rectangular mesh, it is relatively easy to formulate the quadrilat-
eral polygonal mesh of the surface in $\mathbf{R}^3$. Thus, the mapping, which exists between
the (u, v) points in the parameter space and the points in $\mathbf{R}^3$ of the surface, is used
to create an ordered polygonal set of surface points in $\mathbf{R}^3$. Figure 5.3(a) shows the
PDE surface which corresponds to the boundary curves shown in Fig. 5.3(b).

As an example of interactive surface manipulation, let us take the initial surface
shown in Fig. 5.4(b) as the initial surface with associated boundary curves shown in
Fig. 5.4(a). The influence of the derivative vector field on the shape of the surface
can be seen by considering Fig. 5.4(c) and Fig. 5.4(d). In Fig. 5.4(c), the derivative
curve associated with $u = 1$ has been translated 'downwards' along the direction
of local $\mathbf{X}_u$. This changes the position of the derivative curve, and hence a change
in the size of the derivative vector field is brought about. Figure 5.4(d) shows the
surface corresponding to the boundary conditions shown in Fig. 5.4(c).

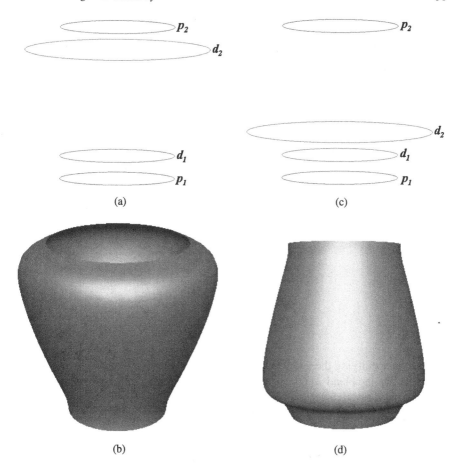

Fig. 5.4 The effect of change in the size and the direction of the derivative vector on the shape of a Biharmonic PDE surface

5.2 Trimming PDE Geometry

For many practical designs, it may be necessary to blend one surface, or part of it, to another. In blending, we are often given the 'primary surfaces' which define the bulk of the object's shape, and we seek 'secondary surfaces' to form a smooth transition between the primary surfaces [5]. The problem of generating a smooth blend between two given adjacent surfaces is not a trivial task, e.g. in many cases, to obtain a satisfactory blend, some form of 'trimming' of the associated surface needs to be performed.

As far as surface design using PDEs is concerned, the problem of blend gener-ation is thought of as a boundary-value problem in which a blending PDE surface patch is generated between the boundaries defined on the adjacent surfaces. Since the boundaries of the two surfaces, between which the blend needs to be generated, are well-defined, it is straightforward to generate a surface patch between the two

Fig. 5.5 Identification of a
point on the parameter space
and the corresponding point
on the surface

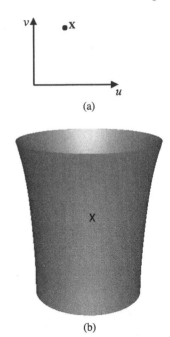

(a)

(b)

surfaces by specifying the appropriate derivative conditions. If tangent-plane continuity between the blend and the surface it meets is required, this can be imposed by
choosing the derivative vectors appropriately.

In order to trim a region of the surfaces, one would first need to determine the
exact position of the surface from which the portion needs to be removed. Any
chosen point in the (u, v) parameter space will have an associated point on the
surface. For the purpose of removing the section interactively, the use of a graphical
interface associated with the interactive environment will enable a user to visualize
both the surface and the (u, v) parameter space. For example, in Fig. 5.5(b), the
point marked '**X**' on the surface is identified as the image point of the marked (u, v)
shown in Fig. 5.5(a) in the parameter space.

Furthermore, the image of a plane curve drawn in the (u, v) parameter space
will be guaranteed to lie on the surface. This curve can then be manipulated in the
parameter space—in the case of a polynomial B-spline, for example, by moving its
control points to achieve the desired shape in $\mathbf{R}^3$ of the portion of the surface to be
removed. The image curve on the surface can then be interactively manipulated by
moving the control points of the curve in the (u, v) parameter space since the shapes
of both curves can be visualized simultaneously. Figure 5.6(a) shows a curve in the
parameter space and Fig. 5.6(b) shows the image of the same curve on the surface.
Thus, it is not a difficult task to create the desired shape of a curve on the surface
by editing the curve in the parameter space in the light of real time feedback on the
shape of the surface curve from its image on the screen.

Once the shape of the curve in the parameter space is decided, a rectangle bounding the curve in the parameter space is determined. Note that the edges of the rect-

Fig. 5.6 A curve on the parameter space and the corresponding curve in $\mathbf{R}^3$ on the surface

(a)

(b)

Fig. 5.7 A curve and its bounding rectangle in the (u, v) parameter space

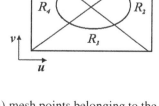

angle lie along the parameter lines. The original (u, v) mesh points belonging to the interior of this rectangle are then discarded. A separate (u, v) mesh is then calculated over the annular region between the rectangle and the curve in the parameter space, which amounts to a re-parametrization of the original surface over its annular region.

One way to re-parameterize the trim region to determine the new (u, v) points is by means of a bilinear interpolation between points on the rectangle and the associated points on the curve as described below.

Figure 5.7 shows a curve in the (u, v) parameter space and the bounding rectangle. It is required to calculate a mesh of points for the region R between the curve and the rectangle.

Using known points on the curve, one can determine the centroid (C_u, C_v) of the curve. The next step is to define a local Cartesian coordinate system in $\mathbf{R}^2$ with the point (C_u, C_v) being at the origin. Here the local x and y axes are taken to be

Fig. 5.8 The (u, v)
parameter space

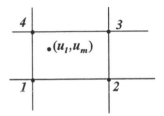

parallel to the u and v directions of the parameter space, respectively. The points on
the curve which are to be associated with the four corners of the rectangle can now
be determined from the intersection of the line segment between (C_u, C_v) and the
corresponding corner of the rectangle. This allows the region R to be subdivided
into four regions R_1, R_2, R_3 and R_4 as shown in Fig. 5.7 and Fig. 5.8.

Bilinear interpolations can now be carried out for each of these four regions sep-
arately. Since the same procedure can be used to compute the bilinear interpolation
for each of the regions R_1, R_2, R_3 and R_4, here it is shown how to compute such
an interpolation for the region R_1. Similar computations can be performed for the
regions R_2, R_3 and R_4.

Assuming that the number of discrete points on the curve segment s_1s_2 is p,
we can subdivide the line segment r_1r_2 of the rectangle, uniformly with distance,
$(p-1)$ times. This provides us with points on the curve segment s_1s_2 and associated
points on the line segment of the rectangle r_1r_2. Now we subdivide each of the line
segments, formed between the points on the curve segment s_1s_2 and the associated
points on the line segment r_1r_2, $(p-1)$ times. This provides us with a grid of $p \times p$
points within the region R_1. The procedure can then be repeated for the rest of the
regions R_2, R_3 and R_4 to obtain the complete mesh.

Once the mesh between annular region of the curve and the bounding rectangle is calculated, it is then straightforward to calculate the corresponding surface
points and surface normals for this new parametrization of the original surface using
the routines which computed these quantities for the original surface. Figure 5.9(a)
shows a wire-frame version of a surface from which a portion has been removed.
Figure 5.9(b) shows the same surface from which a number of other regions of different shapes and sizes have been removed interactively.

The calculation of the mesh over the annular region between the curve in the
parameter space and the bounding rectangle depends on the shape of the curve. For
a curve having a convex shape, the mesh can be calculated by bilinear interpolation
between points on the rectangle and the associated points on the curve as described
above. However, one has to be more careful where a portion of the curve is con-
cave. One possible method to avoid this sort of complications is to perform complex
meshing on trim regions of (u, v) and to use the Laplace equation involving the trim
curves over the (u, v) parameter space as discussed below.

Consider Fig. 5.10 showing the parametric region where the circular curve
marked as C_1 is a trim curve. In order to define a valid mesh within the interior
of the parametric region which discards the trim region, one can solve the Laplace
equation subject to a set of two boundary conditions. Assuming the region in which

Fig. 5.9 Examples of
sections removed from PDE
surfaces

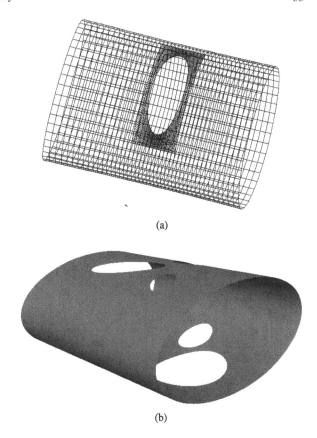

(a)

(b)

meshing is to be performed can be mapped to a finitely discretized region belonging
to $0 \le u \le 1$ and $0 \le v \le 2\pi$, one can assume the mesh $\mathbf{M}(u, v)$ is the solution of
the Laplace equation

$$\left(\frac{\partial^2}{\partial u^2} + \frac{\partial^2}{\partial v^2}\right)\mathbf{M}(u, v) = 0. \tag{5.3}$$

The boundary conditions for Eq. (5.3) are taken to be the curves C_1 and C_2 where
C_2 is the square border curve. Note that both curves C_1 and C_2 are periodic with
$v = 0$ being at the positions marked by the dotted line in Fig. 5.10(a). Thus, one
could imagine that, by solving Eq. (5.3) between the region defined by the curves
C_1 and C_2, we can obtain a smooth transition between the two curves which will
enable us to obtain the required mesh.

In order to carry out the trimming fast enough to be able to generate trimmed
geometry in real time, we utilize analytic solution scheme where the above defined
periodic boundary curves are utilized. This solution scheme is, in fact, very similar
to the analytic solution scheme for the general elliptic PDE discussed previously.
Thus, for Eq. (5.3), the periodic boundary conditions can be expressed as $\mathbf{M}(0, v) =$
$C_1(v)$, $\mathbf{M}(1, v) = C_2(v)$. Figure 5.10(b) shows a surface on which the trim region
corresponding to Fig. 5.10 has been removed.

Fig. 5.10 Meshing the trim region using the Laplace equation. (**a**) A trim curve defined on the parametric domain. The boundary conditions defined by C_1 and C_2 are utilized to solve the Laplace equation within the parametric domain. (**b**) The trimmed PDE surface

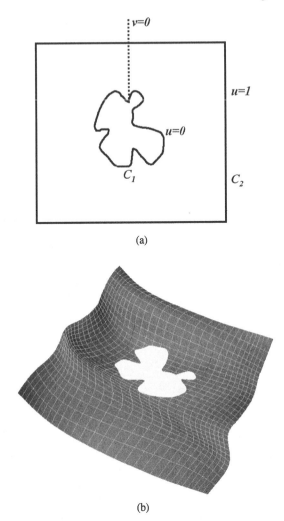

(a)

(b)

5.2.1 Manipulating Blend Geometry

Here we describe, with the aid of an example, how blend geometry can be manipulated. Figure 5.11(a) shows part of a teacup created interactively in which the bowl and the handle are two surface patches blended together, by means of performing appropriate trimming where the handle meets the bowl. Since any point on a surface corresponds to a point in its (u, v) parameter space, the curves where the handle meets the bowl are the images of curves in the (u, v) parameter space of the bowl.

This curve can be changed interactively within the (u, v) parameter space. This results in the translation of the intersection curve on the bowl surface. This procedure allows the position of the handle to be quickly changed interactively.

Fig. 5.11 The shape of an
interactively designed tea cup
and the shape of its
interactively manipulated
handle

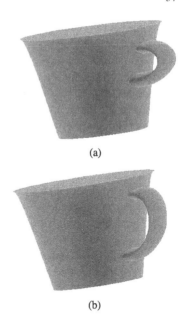

(a)

(b)

Figure 5.11(b) shows the adjustments made to the position of one of the ends of the
handle of the cup starting from that shown in Fig. 5.11(a).

5.3   Spine of PDE Geometry

The spine of an object can be defined as the trace of the centers of all spheres (disks
in the case of two dimensions) that are maximally inscribed in that object. The spine
of an object has a very close geometric resemblance to the more widely known
shape entity called the medial axis or the skeleton. The spine of a shape brings out
the symmetries in that shape and can posses far richer topologies than the shape
it is derived from. Other important properties of the spine of a shape include its
use in the intermediate representation of the object and its canonical general form
that can be used to represent the object by a lower-dimensional description. Thus,
the spine of an object has its natural intuitive appeal in applications in geometric
manipulations. For example, Blum [1] suggested that the spine or the skeleton as
a powerful mechanism for representing the shape of two-dimensional objects at
a level higher than cell-enumeration. He proposed a technique that can uniquely
decompose a shape into a collection of sub-objects that can be readily identified
with a set of basic primitive shapes.

For PDE surfaces described here, the spine of the resulting geometry can be
realized easily by exploiting the structural form of a closed form solution for the
chosen PDE. Thus one can show that the spine of a PDE surface can be generated
as a by-product of this solution [2].

Let us take the general Nth order elliptic PDE described in the previous chapter. If we draw our attention to the term $\mathbf{A}_0$ in Eq. (4.12), we note that it takes the form of a polynomial of degree $(2N - 1)$. By looking at the structure of the PDE solutions given in Eq. (4.11), we can arrive at the following conclusions.

From the structure of the solution to $(2N)$th order PDE given in Eq. (4.11), one can see that a surface point $\mathbf{X}(u, v)$ can be thought of being a sum of the vector $\mathbf{A}_0$ giving the 'center line' of the surface and a radius vector defined by the term $\sum_{n=1}^{\infty}[\mathbf{A}_n(u)\cos(nv) + \mathbf{B}_n(u)\sin(nv)]$ providing the position of $\mathbf{X}(u, v)$ relative to the 'center line'. Thus, the polynomial in the $\mathbf{A}_0$ term describes the spine or the skeleton of the PDE surface [3].

From the very definition of the spine, it can be seen as a powerful and intuitive mechanism to manipulate the shape of surface once it is defined. There are several ways by which one could utilize the spine to manipulate a given PDE geometry. One such possibility is to re-parameterize the spine in terms of a polynomial.

For example, if we take the Biharmonic PDE, we can express the cubic polynomial describing the $\mathbf{A}_0$ term as a Hermite curve of the form

$$\mathbf{H}(u) = \mathbf{B}_1(u)\mathbf{p}_0 + \mathbf{B}_2(u)\mathbf{p}_1 + \mathbf{B}_3(u)\mathbf{v}_0 + \mathbf{B}_4(u)\mathbf{v}_1, \tag{5.4}$$

where the $\mathbf{B}_i$ are the Hermite basis functions, $\mathbf{p}_0, \mathbf{p}_1$ and $\mathbf{v}_0, \mathbf{v}_1$ define the position and the speed of the Hermite curve at $u = 0$ and $u = 1$, respectively. By comparing the Hermite curve given in Eq. (5.4) with the cubic for the spine given by the $\mathbf{A}_0$ term, the terms $\mathbf{a}_{00}, \mathbf{a}_{01}, \mathbf{a}_{02}$ and $\mathbf{a}_{03}$ can be related to the position vectors and their derivatives at the end points of the spine as

$$\mathbf{a}_{00} = \mathbf{p}_0, \tag{5.5}$$

$$\mathbf{a}_{01} = 3\mathbf{p}_1 - \mathbf{v}_1 - 3\mathbf{v}_0, \tag{5.6}$$

$$\mathbf{a}_{02} = \mathbf{v}_1 + 2\mathbf{v}_0 - 2\mathbf{p}_1, \tag{5.7}$$

$$\mathbf{a}_{02} = \mathbf{v}_0. \tag{5.8}$$

Since the $\mathbf{A}_0$ term is an integral part of the solution that generates the surface shape, any change in the shape of the spine will, of course, result in a change in the shape of the surface. A useful mechanism to change the shape of the spine would be to manipulate its position and the derivative at the two end points. Therefore, the position vectors and their derivatives at the end points of the spine can be used as shape parameters to manipulate the shape.

Another way of manipulating the shape is to consider the condiments of the polynomial $\mathbf{A}_0$, e.g., for the general elliptic PDE, one can imagine the vector constants $\mathbf{a}_{00}, \mathbf{a}_{01}, \mathbf{a}_{02}, \ldots, \mathbf{a}_{2N-1}$, as a set of design parameters whose values can be interactively changed.

Figure 5.12(a) shows the spine of corresponding to the PDE surface patch shown in Fig. 5.12(b). Figure 5.13(b) shows a modified shape using the spine, whereby treating it as a cubic spline curve and then manipulating its control points. The final shape of the spine corresponding to this modified geometry is shown in Fig. 5.12(a).

Fig. 5.12 Spine of PDE
geometry. (**a**) The PDE spine.
(**b**) The PDE geometry

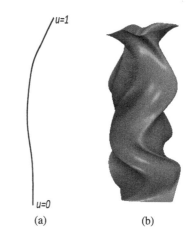

(a)   (b)

Fig. 5.13 Spine based
geometry manipulation.
(**a**) The interactively
manipulated spine from that
shown in Fig. 5.12(a). (**b**) The
resulting PDE geometry

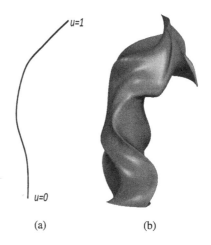

(a)   (b)

5.4 Conclusions

In this chapter, we have discussed practical techniques through which the user is
able to carry out simple surface manipulation interactively in real time. Boundary
conditions for the chosen PDE can be defined in terms of space curves whose shape
then defines the shape of the resulting surface. Another useful interactive tool is the
use of the spine of a PDE surface which also enables the surface to be manipulated in
an intuitive fashion. Owing to the closed-form nature of the solution, its calculation
and recalculation is computationally very efficient, i.e. the surface change produced
by the alteration of any boundary conditions by the user is very rapid. This enhances
real time interactive manipulation of the geometry of complex shapes.

References

1. Blum H (1976) A transformation for extracting new descriptors of shape. In: Wathen-Dunn W (ed) Models for perception of speech and visual form. MIT Press, Cambridge, pp 362–381
2. Ugail H (2003) On the spine of a PDE surface. In: Wilson MJ, Martin RR (eds) Mathematics of Surfaces X. Springer, Berlin, pp 366–376
3. Ugail H (2004) Spine based shape parameterisation for PDE surfaces. Computing 72:195–206. doi:10.1007/s00607-003-0057-8
4. Ugail H, Bloor MI, Wilson MJ (1998) On interactive design using the PDE method. In: Mathematical methods for curves and surfaces II. Vanderbilt University Press, Nashville, pp 493–500. ISBN:0-8265-1315-8
5. Woodward CD (1987) Blends in geometric modelling. In: Martin RR (ed) Mathematical methods of surfaces II. Oxford University Press, London, pp 255–297

Chapter 6
Parametric Design

Abstract This chapter provides details of how PDE based geometries can be efficiently parameterized. The geometry generated using PDEs has an efficient parametrization associated with it. That is, PDE based geometries are first of all characterized by boundary conditions. Furthermore, one can change the geometry easily by means of changing a small set of parameters.

6.1 Design Parameters via the Boundary Curves

In the previous chapter, we saw how an initial surface shape can be defined through boundary curves and then globally manipulated interactively. Here we show how the shape can be further manipulated through design parameters, in particular design parameters that can be introduced on the boundary curves [1, 2, 5, 7].

The basic idea here is to create the boundary curves by defining them in terms of cubic B-splines or using standard analytic descriptions such as circles and ellipses. Thus, by solving the PDE, an initial shape is created which may be an initial approximation to what the user intends to create. In order to facilitate interactive manipulation, a parametrization of the boundary curves can be introduced. Furthermore, the parameters themselves have an obvious 'physical' interpretation and the surface responds in an intuitive manner to the change in the values of these parameters.

Consider the curve shown in Fig. 6.1. Now the points marked X_1 and X_2 can be chosen interactively on the curve. The position of the point X_1 in $\mathbf{R}^3$, the angle θ the line $X_1 X_2$ makes with respect to a given direction and the Euclidean distance s between the points X_1 and X_2 define three parameters which we can refer to as the 'position parameter', the 'angling parameter' θ and the 'scaling parameter' s, respectively. The roles of these parameters are as their names suggest: the position parameter allows the position of the entire curve to be changed in $\mathbf{R}^3$ by 'dragging' the point X_1; the angling parameter θ allows the curve to be rotated about a given direction through the point X_1, and the scaling parameter s allows a given curve to be resized.

Note that the above described procedures can be carried out via simple geometric transformations. In the case of changing the position parameter, this is carried out by translating the point X_1 in $\mathbf{R}^3$. The associated curve is allowed to follow the point,

H. Ugail, *Partial Differential Equations for Geometric Design*,
DOI 10.1007/978-0-85729-784-6_6, © Springer-Verlag London Limited 2011

Fig. 6.1 Illustration of
locally defined parameters
from the boundary curves

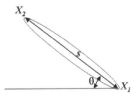

thus translating the whole curve with respect to the value of the parameter X_1. The translations can be carried out in the xy, xz and yz planes of $\mathbf{R}^3$ interactively. In the case of changing the angling parameter θ, a plane (xy, xz or yz) is chosen in which the axis of rotation is to lie. One can then define a line through the point X_1 in this plane. The curve is then rotated about this chosen line through the point X_1 on the chosen plane. Finally, for the scaling parameter, the point X_2 on the curve is chosen and, in order to change the size of the curve, the Euclidean distance between the points X_1 and X_2 is varied.

Thus, by choosing the above model for the geometric transformation of the boundary curves, a set of design parameters are defined influencing the shape of the curve which is easy for the user to appreciate. One can see that despite having kept the geometric transformations as simple as possible, in order to minimize the number of design parameters, the user is able to create a wide range of modifications to the initial curve shape.

Described below is how the above defined parameters can be varied interactively in order to bring about changes in the shape of the surface.

Since the positional boundary conditions define the edge of the surface patch, any change in these conditions directly affects the shape of the surface. Figure 6.2 and Fig. 6.3 show a sequence of surfaces which illustrates the influence of the positional boundary conditions on the shape of the surface where the boundary conditions are controlled by the choice of values for the parameters X_1, θ and s. Here we show how the interactive design of a swirl port [4] for a diesel internal combustion engine can be carried out, starting from an approximation to the desired surface shape shown in Fig. 6.2(b). This initial surface is somewhat 'distant' from the desired final shape of the surface and has been chosen to illustrate the effect of the shape parameters introduced above.

Consider the surface shape in Fig. 6.2(b), which corresponds to the initial boundary curves shown in Fig. 6.2(a). The surface is then modified by changing the boundary conditions, and Fig. 6.2(d) shows the surface resulting from the modified boundary curves which are shown in Fig. 6.2(c). This change to the surface is brought about by changing the value of the position parameter for the curve marked P_2, so that between Fig. 6.2(a) and Fig. 6.2(c) the curve P_2 has been translated in $\mathbf{R}^3$.

The surface is changed again, and Fig. 6.3(b) shows the surface resulting from the modified boundary curves shown in Fig. 6.3(a). The change to the surface shown in Fig. 6.3(b) is brought about by changing the value of the angling parameter for the curve marked P_2 shown in Fig. 6.2(c).

Figure 6.3(d) shows the final shape of the swirl port. The corresponding boundary curves are shown in Fig. 6.2(c). The change to the surface shown in Fig. 6.2(b)

Fig. 6.2 The influence of
positional boundary
conditions on the shape of the
surface

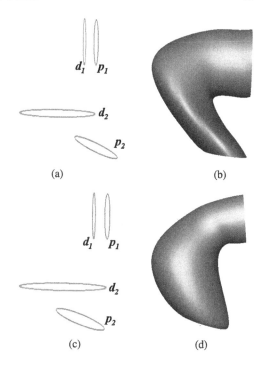

(a)   (b)

(c)   (d)

results from increasing the value of the scaling parameter s for the curve marked P_2
shown in Fig. 6.2(a).

6.2 Local Parameters on the Boundary Curves

As shown in the previous section, the boundary conditions are very powerful in
controlling the internal shape of the surface. For this reason, modification of the
boundary curves themselves at a local level may often be required. Described below
is a technique which has been implemented to achieve this, using a locally defined
parametrization.

Next, the local modification of a B-spline curve is considered, although in principle it can be any sort of parametric curve. Consider the curve shown in Fig. 6.4(a).
Two points X_1 and X_2 on the curve are chosen; between these points the user seeks
to manipulate the curve. Since the curve is arc-length parameterized in terms of v,
the points X_1 and X_2 on this parametrization correspond to, say, v_1 and v_2 with
$v_1 \leq v_2$. We define a new curve $\underline{C}_{new}(v)$ such that

$$\underline{C}_{new}(v) = \begin{cases} \underline{C}_1(v) & \text{if } v \leq v_1, v \geq v_2, \\ \underline{C}_2(v) & \text{if } v_1 < v < v_2, \end{cases} \qquad (6.1)$$

Fig. 6.3 The influence of
positional boundary
conditions on the shape of the
surface

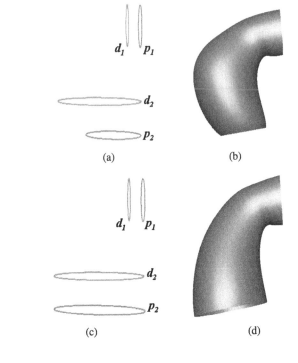

(a)   (b)

(c)   (d)

Fig. 6.4 Illustration of how
the boundary curves can be
manipulated locally

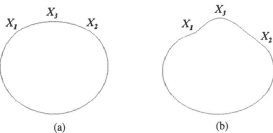

(a)   (b)

where $\underline{C}_1(v)$ is the original curve and $\underline{C}_2(v)$ is a B-spline perturbation curve given
by

$$\underline{C}_2(v) = \sum_i \underline{P}_i B_i(v), \tag{6.2}$$

where B_i is a cubic B-spline and $\underline{P}_i$ are the control points. The knot values of the
B-spline are chosen so as to confine the effects of manipulation to a given region
of the original curve. In particular, the point X_3, which is initially on the curve,
corresponds to a control point of the B-spline curve, and it is interactively chosen
between X_1 and X_2. To ensure that the curve $\underline{C}_2(v)$ passes through the end points
of the curve $\underline{C}_1(v)$ and at the same time maintains the continuity between the two
curves, the following procedure has been employed.

Fig. 6.5 Influence of local manipulation of the derivative curve on the shape of the surface

 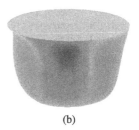

(a)   (b)

Assuming that the points on the original curve are labeled as v_i, with increasing i as the points on the curve are counted from X_1 to X_2, we can define the point X_1 to be v_m for some m. Similarly, X_2 is defined to be v_n with $n > m$. The points v_{n-1} and v_{m+1} are now defined as multiple control points (v_{n-1} and v_{m+1} each as three control points) of the B-spline curve. This ensures that the curve $\underline{C}_2(v)$ passes through the points v_{n-1} and v_{m+1}. This also ensures that the slopes at the end points of the curve $\underline{C}_2(v)$ are the vectors $(v_{n-1} - v_n)$ and $(v_{m+1} - v_m)$.

Since the aim of introducing the curve $\underline{C}_2(v)$ is to enable local manipulation, the point X_3, which the user is able to choose between X_1 and X_2 interactively, is the only point which is allowed to be manipulated. Thus, X_3 is allowed to be translated in the plane defined by the initial choice of points X_1, X_2 and X_3. Figure 6.4(b) shows the resulting curve after a local manipulation has been applied to the curve shown in Fig. 6.4(a).

The above described procedure has also been employed to create curves in $\mathbf{R}^3$ from planar curves. In order to do this, the point X_3 once chosen can be translated along the line through X_3 which is perpendicular to the plane defined by X_1, X_2, and X_3.

Thus, parameters X_1, X_2 and X_3, which are chosen interactively on the curve by input from the mouse, can be varied to change the shape of the curve locally. This local modification in the curve readily gives rise to a complex modification of the surface.

Figure 6.5(b) shows a cup-like surface to which a 'lip' has been added in which the derivative curve at $u = 1$ has been locally manipulated in the fashion described above. The initial surface patch is shown in Fig. 6.5(a).

6.3 The Effect of the Smoothing Parameter a

The parameter a given in the general Biharmonic equation (4.35) also influences the overall shape of the surface. This parameter controls the relative smoothing of the dependent variables between the u and v directions. For large a, changes in the u direction occur over a relatively short length scale, i.e. it is $1/a$ times the length scale in the v direction over which similar changes take place. Thus, by adjusting the value of a interactively, the user is capable of controlling the length scale over which the boundary conditions influence the interior of the surface. For a periodic surface, the higher the value of a, the more 'waist' the surface acquires. Figure 6.6

Table 6.1 Values of *a* used to create the geometries shown in Fig. 6.6	Fig.	Value of *a*
	6.6(a)	1.5000
	6.6(b)	5.7500
	6.6(c)	8.1800

shows the sequence of surfaces resulting in changes to the value of a, for a cup-like surface in the case of the Biharmonic equation. The particular values of a chosen are shown in Table 6.1.

Figure 6.6 shows a series of geometry created using various values for the parameter a. Note that as far as this work is concerned, the parameter a is chosen to be independent of u and v. In certain specialized cases, it may be thought desirable for this parameter to be made dependent on u, v so that more control over the shape of the surface can be obtained [3].

6.4 The Effect of v Parametrization

The points on the boundary curves where $v = 0$ are important in defining the overall shape of the surface. As described earlier, during the initial design process, the position of this point on each of the boundary curves can be chosen interactively. The effect of changing the $v = 0$ point from its original position to another position on a given curve is to produce a twist in the parameter lines.

Consider Fig. 6.7(a) in which the point $v = 0$ at $u = 1$ is marked. Figure 6.7(b) shows the result of changing the position of this point at $u = 1$ on the boundary curves corresponding to surface shown in Fig. 6.7(a).

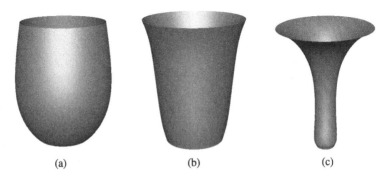

(a)   (b)   (c)

Fig. 6.6 Interactive shape manipulation using the parameter a

Fig. 6.7 The effect of v parametrization

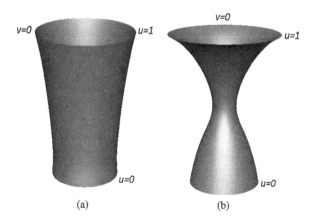

(a)   (b)

6.4.1 Time-Dependent Parametrization

The use of time as a parameter (especially on the boundary conditions) can be used as a design parameter. To describe this, we show how shape morphing can be carried out by means of introducing a time parameter to the boundary conditions.

Let $\mathbf{B_s} = \{\mathbf{S_1}, \mathbf{S_2}, \mathbf{S_3}, \mathbf{S_4}\}$ be the set of boundary conditions specified for the $\mathbf{S_s}$ and $\mathbf{B_t} = \{\mathbf{T_1}, \mathbf{T_2}, \mathbf{T_3}, \mathbf{T_4}\}$ be the set of boundary conditions representing $\mathbf{S_t}$. Now, the ith intermediate set of boundary conditions $\mathbf{B_i} = \{\mathbf{I_1}, \mathbf{I_2}, \mathbf{I_3}, \mathbf{I_4}\}$ can be achieved by

$$I_1 = (1 - \varepsilon)\mathbf{S_1} + \varepsilon\mathbf{T_1},$$
$$I_2 = (1 - \varepsilon)\mathbf{S_2} + \varepsilon\mathbf{T_2},$$
$$I_3 = (1 - \varepsilon)\mathbf{S_3} + \varepsilon\mathbf{T_3},$$
$$I_4 = (1 - \varepsilon)\mathbf{S_4} + \varepsilon\mathbf{T_4},$$

(6.3)

where

$$\varepsilon = \frac{\gamma i}{m}$$

(6.4)

with $\gamma \geq 0$ and m being the total number of intermediate surfaces to be created.

Figure 6.8(a) corresponds to the water glass which is transformed into Fig. 6.8(b) by changing its boundary conditions when $i = 1$. Then, Fig. 6.8(b) evolves into Fig. 6.8(c) and successively to Fig. 6.8(k) which represents a wine glass.

6.5 Summary

This chapter has highlighted the aspects of parametric design using PDEs. It has been shown that PDE based geometric design represents a powerful means for parameterizing complex geometries due to its inherent boundary-value approach to

Fig. 6.8 Sequence showing how the source surface $\mathbf{S_s}$ has been morphed into the target one $\mathbf{S_t}$ by using a gradual change in the boundary conditions and finding each of the intermediate surfaces

surface generation. Then a number of different types of parametrization have been discussed. In particular, shape parameters introduced to the boundary curves as well as time-dependent parameterizations have been considered.

References

1. Athan M, Ugail H, Gonzalez G (2009) Parametric design of aircraft geometry using partial differential equations. Adv Eng Softw 40:479–486. doi:10.1016/j.advengsoft.2008.08.001
2. Brown JM (1992) The design and properties of surfaces generated using partial differential equations. PhD thesis, Department of Applied Mathematical Studies, University of Leeds, UK
3. Cheng SY (1992) Blending between parametric surfaces using partial differential equations. PhD thesis, Department of Applied Mathematical Studies, University of Leeds, UK
4. Dekanski CW, Bloor MIG, Wilson MJ (1996) A parametric model of a two-stroke engine for design and analysis. Comput Methods Appl Mech Eng 137:411–425. doi:10.1016/S0045-7825(96)01103-6

5. Kubeisa S, Ugail H, Wilson M (2004) Interactive design using higher order PDEs. Vis Comput 20(10):682–693. doi:10.1007/s00371-004-0261-3
6. Ugail H (2004) Spine based shape parameterisation for PDE surfaces. Computing 72:195–206. doi:10.1007/s00607-003-0057-8
7. Ugail H, Bloor MI, Wilson MJ (1999) Manipulation of PDE surfaces using an interactively defined parameterisation. Comput Graph 23(4):525–534. doi:10.1016/S0097-8493(99)00071-0
8. Ugail H, Bloor MIG, Wilson MJ (1999) Techniques for interactive design using the PDE method. ACM Trans Graph 18(2):195–212. doi:10.1145/318009.318078

Chapter 7
Functional Design

Abstract This chapter discusses how the PDE based approach to shape parametrization when combined with a standard method for numerical optimization is capable of setting up automatic design optimization problems allowing design for function to be more practical. The chapter first introduces the methodologies for design optimization. Afterwards several examples of how design for function can be carried out via PDE based shape parametrization along with optimization are discussed.

7.1 Introduction

In the previous two chapters, it was shown how it is possible to change the shape of a parametric PDE surface by interactively changing the values of the previously defined parameters. Thus, the user can search by eye, varying the parameters and viewing the resulting surface graphically, until the desired shape is achieved. However, it is often necessary to obtain surfaces which possess some functional property. For example, it may be necessary to search for a surface, among those available, which encloses the largest volume, or which has a particular curvature distribution. To perform such tasks automatically, the surface generation routine needs to be combined with a routine for analysis that evaluates the values of a function, usually termed the objective function, that quantifies the property of interest, and with a further routine, termed the optimization routine, which automatically searches for a minimum (or maximum) of the objective function.

Generally, the optimization process requires a search to be made in the parameter space in order to find the minimum value of the objective function. During the process of optimization, most of the computational effort is usually spent on evaluating the objective function rather than the optimization routine itself. Therefore, it is desirable to use a design method which minimizes the number of design variables and therefore requires as few function evaluations as possible [5].

7.2 Principles of Shape Optimization

The process involved in a standard shape optimization routine is to solve a constrained optimization problem in which an objective function f the design shape

H. Ugail, *Partial Differential Equations for Geometric Design*,   71
DOI 10.1007/978-0-85729-784-6_7, © Springer-Verlag London Limited 2011

parameters and the chain of constraints are related. In the interest of illustrating this, a very simple optimization example is explained. Assume that you are interested in the design of a liquid bottle whereby the container is cylindrical with radius r and height h. Due to cost issues, you require to keep the total surface area of the container fixed to 3π square meters. However, you need to maximize the volume of the container in order to maximize the profit. This is clearly an optimization problem in which the volume of the cylinder needs to be maximized subject to the constraint stating that the area must remain constant. Let r and h denote the radius of such a cylinder and let A and V represent the total surface area and volume of such a cylinder, which are given by

$$A = 2\pi r^2 + 2\pi rh$$

and

$$V = \pi r^2 h,$$

respectively.

The total surface area includes two circular sections comprising the top and the base of the cylinder. Here the volume of the cylinder is the objective function, and we will denote it by f. In this particular case, f is a function of both the radius and the height of the cylinder. The constraint is naturally given by the restriction imposed by limiting the area of material to be 3π square meters. This can be written as

$$A = 3\pi = 2\pi r^2 + 2\pi rh.$$

This relation can be exploited in order to find a relation between the two variables defining the geometry of the cylinder r and h, leading to

$$h = \frac{3}{2r} - r.$$

The next step is to express the objective function f exclusively in terms of r, obtaining

$$f = \frac{3}{2}\pi r - \pi r^3.$$

Once the objective function has been expressed in terms of one variable, one can proceed to maximize the function. The traditional approach for either maximizing or minimizing a function with respect to a given variable is to find its critical points by differentiating it with respect to the variable to be optimized. Thus,

$$\frac{df}{dr} = \frac{3}{2}\pi - 3\pi r^2 = 0,$$

and the critical points are then

$$r = \frac{1}{\sqrt{2}} \quad \text{and} \quad r = \frac{-1}{\sqrt{2}}.$$

Clearly, the critical point at $r = \frac{-1}{\sqrt{2}}$ is automatically discarded since there cannot be circles of negative radii, and therefore, our interest will be focused on the second

Table 7.1 Values assigned to r and the respective value of h for function f

r	h	f
$\frac{1}{4}$	$\frac{35}{4}$	$\frac{35}{64}\pi$
$\frac{1}{3}$	$\frac{25}{6}$	$\frac{25}{54}\pi$
0.5	$\frac{5}{2}$	$\frac{5}{8}\pi$
$\frac{2}{3}$	$\frac{19}{12}$	$\frac{19}{27}\pi$
$\frac{1}{\sqrt{2}}$	$\sqrt{2}$	$\frac{1}{\sqrt{2}}\pi$
$\frac{\sqrt{3}}{2}$	$\frac{\sqrt{3}}{2}$	$\frac{3\sqrt{3}}{8}\pi$
1.0	0.5	$\frac{1}{2}\pi$

one at $r = \frac{1}{\sqrt{2}}$. Now, it is necessary to verify that such a critical point is indeed a maximum. To that end, a number of mathematical criteria based on derivatives of the objective function are available; however, the reader is referred to Table 7.1 to verify that such a value is a maximum. This table presents different values for r, h and f, which by simple inspection demonstrates that the greatest volume is at $r = \frac{1}{\sqrt{2}}$.

It is important to stress that the example above is simple, and a straightforward analytic solution to the corresponding optimization problem is available. However, this is not generally the case, and more complicated mathematical techniques have to be employed. A number of numerical optimization techniques are readily available. One such method is the so-called Simulated Annealing.

7.3 Simulated Annealing

Simulated annealing is based on an analogy between the simulation of the annealing process in cooling solids and the problem of solving large combinatorial problems. This analogy was first remarked upon in [7]. In condensed matter physics, annealing denotes the physical process in which a solid, or liquid, is heated to high temperatures and then cooled slowly in order to remove strain crystal imperfections. At sufficiently high temperatures, the particles of the melted solid are free to move randomly in the liquid phase. If the temperature is reduced sufficiently slowly, such that the thermal mobility is reduced gradually, the particles tend to arrange themselves to form a crystalline structure of minimum energy. An important property of annealing is that the gradual reduction in temperature allows the solid to reach thermal equilibrium at each temperature value T.

The probability that a solid, in thermal equilibrium, is in a state of energy E is given by the Boltzmann probability distribution function

$$P(E) = \frac{1}{Z(T)} e^{-\frac{E}{kT}}, \tag{7.1}$$

where $Z(T)$ is a normalization factor, known as the partition function, which depends on the temperature T, and k is a constant of nature, known as Boltzmann's constant, which relates temperature to energy. A detailed discussion of this can be found in [8]. According to Eq. (7.1), as the temperature decreases, the Boltzmann distribution concentrates on states with lowest energy, and finally, when the temperature approaches zero, only the minimum energy states have a non-zero probability occurrence.

However, if the cooling is too rapid, the system is unable to reach thermal equilibrium at each temperature value. This can result in defects being 'frozen' into the solid, yielding a polycrystalline structure of considerably higher energy when compared to pure crystalline structure. Furthermore, in a process known in condensed matter physics as quenching, if the temperature is lowered instantaneously, it results in defects being frozen into the solid, resulting in a metastable structure.

An algorithm for simulated annealing was first introduced in [9]. In that algorithm, the authors simulated the evolution of a solid into a thermal equilibrium at a given temperature. The algorithm is based on the Monte Carlo method, which is used to estimate averages or integrals by means of random sampling techniques. The algorithm is used to generate a sequence of states of the solid in the following way.

In each step of the algorithm, a particle was given a small randomly generated perturbation and the resulting change in energy ΔE of the system was calculated. If the perturbation results in a lower energy, i.e. $\Delta E < 0$, the new state is automatically accepted. However, if the perturbation results in a higher energy, i.e. $\Delta E \geq 0$, then the probability of accepting the new state is given by the Boltzmann's factor $P = e^{\frac{-\Delta E}{kT}}$. In order to decide whether to accept the proposed new state, a random number is generated in the interval $(0, 1)$, and if P is greater than this random number, the new state is accepted. This acceptance rule for a new state is referred to as the Metropolis criterion, and the general scheme of always accepting a lower state and sometimes accepting a higher energy state is known as the Metropolis algorithm. After a large number of perturbations, using the Metropolis criterion, the probability distribution approaches that of the Boltzmann distribution, i.e. thermal equilibrium.

The Metropolis algorithm was generalized by [7] by the introduction of an 'annealing schedule' in order to solve arbitrary combinatorial minimization problems. This algorithm is known as simulated annealing, and is really a sequence of Metropolis algorithms performed at a sequence of decreasing temperature values, i.e. at each temperature value the Metropolis algorithm is performed until thermal equilibrium is achieved, and, provided that the annealing schedule is sufficiently slow, simulated annealing converges to a global minimum corresponding to the minimum energy state of the system. It is the controlled acceptance of the perturbations which enables simulated annealing to jump from a local minimum into a potentially better, and hopefully global, minimum.

Simulated annealing was originally devised for the solution of combinatorial optimization problems which involve a discrete, and often very large, configuration space. The algorithm has been applied to a variety of problems, including the traveling salesman problem and computer circuit design. The basic idea of simulated annealing is also applicable to problems with continuous search spaces, e.g. the non-linear least squares fitting problem [1]. Advantages of simulated annealing over local minimization methods include its general applicability, flexibility, robustness, and ease of implementation. A disadvantage of the simulated annealing method is that it involves more function evaluations than gradient-based methods.

7.4   Application of Simulated Annealing to Continuous Optimization Problems

Described in this section is how simulated annealing has been implemented to solve the optimization problem. In particular, it is shown how the algorithm is formulated and how the automatic variation of the geometry is carried out.

7.4.1   Simulated Annealing Algorithm

The algorithm used for this work is based on the Metropolis algorithm and is used to generate a sequence of configurations of the search space. We let configurations play the role of energy states of a solid, while the objective function f and a control parameter c play the role of energy and temperature, respectively. The algorithm then evaluates a sequence of Metropolis algorithms at a sequence of decreasing values of the control parameter. A generation mechanism is then defined, so that, given a configuration i, another configuration j can be obtained by choosing at random an element from the neighborhood of i, the latter corresponding to the small perturbation in the Metropolis algorithm. Let $\Delta f_{ij} = f(j) - f(i)$, then the probability for configuration j to be the next configuration in the sequence is given by 1, if $\Delta f_{ij} \leq 0$ and by $e^{-\frac{\Delta f_{ij}}{c}}$, if $\Delta f_{ij} > 0$ (the Metropolis Criterion). Thus, there is a nonzero probability of continuing with a configuration with higher cost than the current configuration. This process is continued until equilibrium is reached, i.e. until the probability distribution of the configurations approach the Boltzmann distribution, now given by

$$P(\text{configuration} = i) = \frac{1}{Z(c)} e^{-\frac{f(i)}{c}}, \tag{7.2}$$

where $Z(c)$ is a normalization constant depending on the control parameter c.

The control parameter is then lowered in steps, with the system being allowed to approach equilibrium for each step by generating a sequence of configurations. The algorithm is terminated for some small value of c. The final 'frozen' configuration is then taken as the solution of the problem. Note that the acceptance criterion is implemented by drawing random numbers from a uniform distribution on $(0, 1)$ and comparing these with $e^{-\frac{\Delta f_{ij}}{c}}$ as shown below.

The rate at which the system cools can be controlled by introducing an appropriate annealing schedule. This can effectively be done by introducing a scheme to lower the control parameter. Here an exponential annealing schedule is chosen and is given by

$$c_m = c_0 e^{(s-1)m}, \quad 0 < s < 1, \tag{7.3}$$

where c_m is the temperature at any given step m and c_0 is the initial temperature.

The pseudo-code for the algorithm is given below. Note that the algorithm outlines the setup for one design parameter, and the generalization of the algorithm to n design parameters can be carried out in an obvious way.

Algorithm Used for Simulated Annealing
Initialize();
$m = 0$;   // *m introduces an annealing schedule*
while **stop criterion** = **FALSE**
{
   while **equilibrium is approached sufficiently closely** = **FALSE**
   {
   PERTURB(config. $i \rightarrow$ config. j, Δf_{ij});   // *generates new configurations*
   if { $\Delta f_{ij} \leq 0$ then accept   }
   else if { $\exp(-\Delta f_{ij}/c) >$ random$(0, 1)$ then accept   }
   if accept then **UPDATE**(config. j);
   }
   $c_{m+1} = L(c_m)$;   // *calculate a lower value for c*
   $m = m + 1$;   // *define a new equilibrium state*
}
system is 'frozen';

7.4.2 Constraints

Since simulated annealing is insensitive to discontinuities in the objective function, it is relatively simple to impose constraints on the chosen objective function. The simplest way of doing this is to assign a large positive number to the objective function f if a given parameter constraint is violated, in other words, by introducing a crude penalty function. A sufficiently large penalty function added to the objective function will usually ensure that trespassing over the bounds is avoided.

7.5 Further Examples

This section illustrates how practical design for function can be carried out using a PDE formulation. For this purpose, several examples are discussed.

As a simple example, consider the shape shown in Fig. 7.1(b). The corresponding boundary curves are shown in Fig. 7.1(a). Here the aim was to obtain a shape for the

Fig. 7.1 Initial boundary curves and the corresponding PDE surface to be optimized for curvature

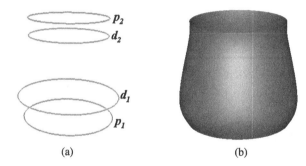

(a)   (b)

surface which has the minimum $\int (\kappa_1^2 + \kappa_2^2)\, ds$, where κ_1 and κ_2 are the principal curvatures, which can be obtained from the well known formulae

$$G = \kappa_1 \kappa_2, \tag{7.4}$$

$$H = \frac{\kappa_1 + \kappa_2}{2}, \tag{7.5}$$

where G and H are known as the Gaussian and mean curvature of the surface, respectively, and are expressed in terms of the first and second fundamental forms of the surface [10].

Here the parametrization discussed in previous chapters is introduced on both the derivative curves d_1 and d_2. The parametrizations are defined using two points (in this case, at $v = 0$ and $v = \pi$) interactively chosen on the curves. For convenience, the parametrizations on the boundary curves are denoted using the notation c_{kP_i} $(k = 1, 2)$, $(i = x, y, z)$. Here c defines the curve, with the letter p being used to denote the position curves and the letter d being used to denote the derivative curves. The index k ranges from 1.0 to 2.0 denoting respectively the $u = 0$ and $u = 1$ boundary edges of the surface. The letter P denotes the type of parameter, T stands for a translation, R for a rotation and D for a dilation. Finally, the letter i denotes the coordinate directions relevant to a particular type of parameter.

For the example considered here, translation, rotation and dilation parameters are defined for both the derivative curves d_1 and d_2 with the position curves kept fixed. In particular, we have translations, d_{kT_i} $(k = 1, 2)$, $(i = x, y, z)$, rotations, d_{kR_i} $(k = 1, 2)$, $(i = x, y, z)$, and dilations, d_{kD_i} $(k = 1, 2)$, $(i = x, y, z)$. Furthermore, we also consider the variation of the smoothing parameter a.

Table 7.2 shows the initial values along with the parameter ranges are considered. These values were chosen in such a way that their variations lead to a sensible range of shapes. To reduce the cost of optimization, it is usually necessary to limit the search regions of the parameter space in which the shapes obtained are sensible, though not necessarily optimal. In practice, a designer makes this sort of design judgment when considering the possible design solutions. In other words, the designer does not waste time looking for alternative designs which would obviously not possess the required functionality.

The above parameters are given to the optimization routine which can be integrated with the software which generates a PDE surface to the Biharmonic equation

Table 7.2 Initial values and ranges for the parameters considered in optimizing a surface for curvature

Parameter	Initial value	Minimum value	Maximum value
d_{kT_i} $(k = 1, 2)$, $(i = x, z)$	0.0	−1.2	1.2
d_{1T_y}	0.0	0.0	1.2
d_{2T_y}	0.0	−1.2	0.0
d_kR_i $(k = 1, 2)$, $(i = x, y, z)$	0.0	−0.8	0.8
d_{kD_i} $(k = 1, 2)$, $(i = x, y, z)$	1.0	0.01	3.0
a	0.5	0.01	9.0

Fig. 7.2 Intermediate shapes due to the variation of the parameters during optimization

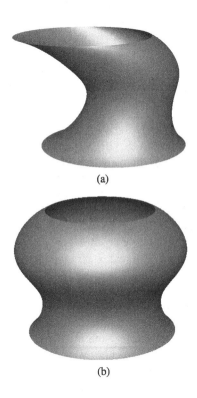

(a)

(b)

with the smoothing parameter a discussed in Chap. 4. Thus, the geometry is varied for each iteration of the simulated annealing algorithm and the corresponding surface points, along with the objective function, are calculated. Figure 7.2 shows some of the intermediate shapes which have been produced during the process of optimization.

Table 7.3 shows the final values of the parameters. Note that the table only shows the values for those parameters whose values changed significantly during the optimization. Figure 7.3 shows the optimal shape of the surface. This shape is obviously a reasonable shape as it has relatively small curvature.

Table 7.3 Final values of the parameters corresponding to the optimal shape for optimal design for curvature

Parameter	Optimal value
d_{1D_x}	0.909
d_{1D_y}	0.818
d_{1D_z}	0.908
d_{2D_x}	1.022
d_{2D_y}	1.015
d_{2D_z}	1.018
a	0.934

Fig. 7.3 Optimal design for curvature

7.5.1 Design Optimization of a Thin-Walled Structure

The aim here is to design a container possessing the minimum amount of material subject to required strength. Moreover, for the purpose of illustration, the problem of stacking the containers on top of each other for the purpose of transportation and display on supermarket shelves is considered. Consider a container at the bottom of a stack, which experiences a stress (due to the weight of the rest of the containers in the stack) and hence becomes slightly deformed. It is the excessive shear stress which can cause most damage to the walls of the container. Thus, a measure for the required strength of the container can be computed by calculating the maximum shear stress [6] within the container. This is done by means of thin-shell finite element analysis [3] where a force is applied around the rim of the container, which translates to the weight of the rest of the containers in the stack. It is also assumed that the base of the container is fixed.

Figure 7.4 shows the shape of a yoghurt container created using two surface patches (one for the bowl and the other for the flat flange). This is a basic shape which can be used to demonstrate the techniques of shape optimization.

The ridges at the base of the container are created so as to obtain a more realistic shape for the container as such features are often incorporated within food contain-

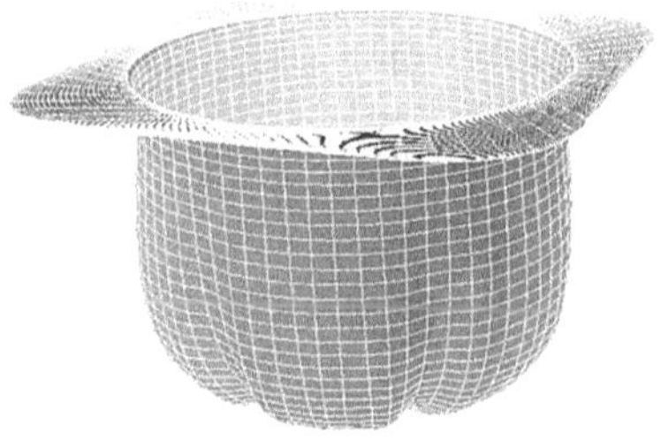

Fig. 7.4 Shape of a yoghurt
container: an example of a
composite shape built from
multiple patches

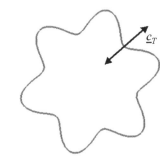

Fig. 7.5 Cubic B-spline
curve corresponding
derivative condition enabling
to create the ridges at the base
of the container

ers. Here these ridges are created by means of the corresponding derivative curve
which in this case is defined by means of a cubic B-spline and takes the form

$$d1(v) = \sum_i \mathbf{c}_i B_i(v), \tag{7.6}$$

where B_i is a cubic B-spline and $\mathbf{c}_i$ are the control points.

The control points, $\mathbf{c}_i$, of the spline are chosen so as to initially lie on the curve
which determines the value of dilation parameter. The number of control points
chosen determines the number of ridges on the container. In order to define the
amplitude of the ridges, one can translate the control points, normal to the curve,
within a confined xy planar region. Thus, the amount of translation of the control
points normal to the curve determines the prominence of the ridges. The translation
of the control points introduces an extra shape parameter which will be referred
as $\mathbf{c}_T$. Figure 7.5 shows the B-spline curve illustrating the parameter $\mathbf{c}_T$.

Here, for convenience, the parametrization on the boundary curves is adopted
to suit the optimization problem using the following notation. For a given boundary curve, this parametrization is denoted as c_{kP_i} $(k = 1, 2)$, $(i = x, y, z)$. Here c
indicates the type of curve, with the letter p denoting the position curves and the
letter d denoting the derivative curves. The index k ranges from 1 to 2 corresponding respectively to the $u = 0$ and $u = 1$ boundary edges of the surface. The letter P
denotes the type of parameter: T stands for a translation, R for a rotation and D for
a dilation. Finally, the index i denotes the coordinate directions relevant to a particular type of the parameter. Adjustments to the values of these parameters along with
the value of a in the Biharmonic PDE can be used to create and manipulate complex
geometries.

Table 7.4 Parameter values for the yoghurt container (optimization for strength)

Parameter	Minimum	Maximum	Initial	Optimal
d_{1T_y}	−0.400	−0.001	−0.400	−0.134
d_{1D_x}	0.100	0.800	0.450	0.300
d_{1D_y}	0.100	0.800	0.450	0.298
d_{2T_y}	0.001	0.400	0.400	0.401
d_{2D_x}	0.100	0.800	0.450	0.370
d_{2D_y}	0.100	0.800	0.450	0.378
a	1.000	7.000	1.000	1.075
c_T	−0.300	0.300	0.200	0.001

As mentioned above, the design objective here is the minimization of the mass of the container subject to a given maximum shear stress. Hence the process of optimization requires the calculation of the maximum shear stress that occurs in the object for every design to be analyzed. Using the principal components of shear stresses, obtained by performing the finite element analysis of the structure, the maximum shear stress σ_{max}^p occurring on any plane through a point p is first calculated. Therefore, the measure for the strength of the container is the maximum shear stress occurring in the whole structure, i.e.

$$f = \underbrace{\max}_{\text{(all points)}} \{\sigma_{max}^p\}. \tag{7.7}$$

Since the determination of the internal shape of the container is of interest here, the only changes in the parameters introduced here are those on the derivative boundary curves of the bowl part of the container. In particular, the translation in y direction and dilations in the xy plane of these two curves within defined limits are considered in order obtain a favorable range of shapes.

With this formulation, the design parameters and their initial values for the optimization of the yoghurt container are shown in Table 7.4. Note that the table also shows the chosen range for each parameter. The range specified for each design parameter (by means of choosing a maximum and a minimum) allows the parameters to be varied within the specified ranges, enabling alternative shapes to be created within the design space automatically. These ranges are chosen to ensure that the geometry of shapes produced within the design space is sensible. The container was loaded so as a tension of 15 $\mathrm{N\,m}^{-1}$ (equivalent to the weight of about 30 yoghurt containers) was applied to the top seal of the container. It is also assumed that the base $u = 0$ of the container is fixed. The required strength of the container is specified so that the level of stress occurring within the loaded structure less than 30% of the yield stress. The design space was further restricted by choosing a volume constraint for the container. For this particular example, a fixed volume of 150 ml was chosen.

Once the geometry is parameterized, the design parameters and their ranges along with the value for the required volume of the container are fed into the opti-

Fig. 7.6 Optimal design for
strength of the yoghurt
container

mization routine. This routine then automatically searches the design space in order
to find the design with the lowest possible value of the chosen merit function.

The values of the parameters obtained for the optimal design are shown in Table 7.4, and the optimal shape is shown in Fig. 7.6. The resulting optimal shape in
this particular case yields a relative reduction in mass of around 11%.

7.5.2 Prediction of Stable Structures of Vesicles Occurring in Biological Organisms

This example discusses how the method of shape parametrization based on the PDE
formulation can be used to predict the stable structures of vesicles commonly found
in biological organisms. The vesicles considered here are essentially lipid molecules
typically consisting of a polar hydrophilic head and a hydrophobic tail consisting of
hydrocarbon chains. Such amphiphilic molecules when placed in an aqueous solution can spontaneously aggregate to form encapsulating bags. Despite the relatively
simple structure of their walls, these vesicles can adopt a surprisingly wide variety
of different shapes and even topologies [4]. It is noteworthy that various shapes and
topologies are adopted by these vesicles during the aggregation so as to reduce the
surface energy of the membrane [2].

Therefore, the aim here is to predict the stable shapes of the vesicles by means
of automatic optimization subject to a given set of physical conditions with the
merit function being the surface energy of the membrane. Various models for predicting the surface energy of a membrane can be found in the literature. For the
work described here, a widely accepted model for predicting the surface energy of
a membrane due to Canham [2] that is based upon the surface curvature (SC) of the
membrane is used. The SC model is based on the fact that the local energy density
of the membrane is proportional to the sum of the squares of the principal curvatures
and a quantity known as the spontaneous curvature to reflect the possible asymmet-
ric configuration of the membrane. According to this model, the shape adopted by a
vesicle is such as to locally minimize the energy functional subject to the constraints
of constant area and volume. Hence, the surface energy E of a vesicle is given by a
surface integral of the form

$$E(S) = \int (C_1 + C_2 - C_0) \, dA + \int (C_1 C_2) \, dA, \qquad (7.8)$$

Fig. 7.7 Initial geometry of the vesicle shape

where C_1 and C_2 are the principal curvatures, C_0 is the spontaneous curvature for a given surface shape S with dA being an element of the surface. Note that if we use the analytic solution of the PDE, the surface is given in closed form, allowing the computation of principal curvatures and the Jacobean relating an area element dA in the (u, v) parameter space.

The approach to predicting vesicle shapes adopted here is similar to that described in the previous example where a parametric representation of the surface corresponding to the shape of a yoghurt container is created. The geometry of the vesicle shape is represented using two PDE surface patches joined together with a common boundary as shown in Fig. 7.7. Once again the design parameters are identified at the boundary curves with their starting values and ranges directly fed to the optimization routine. The optimization is carried out subject to the constraints of constant surface area and enclosed volume. Due to the scale invariance of the SC model, the vesicle shapes depend on two dimensionless parameters known as the reduced volume v and the reduced spontaneous curvature c_0 given by

$$v = V \Big/ \left(\frac{4\pi}{3} R_0^3 \right) \tag{7.9}$$

and

$$c_0 = C_0 R_0, \tag{7.10}$$

where $R_0 = \sqrt{(4\pi A)}$ where V and A are the volume and surface area of the vesicle, respectively.

With these settings, the optimization is started at some initially chosen point in the parameter space. The routine allows detecting the local minimum of the surface energy for a given value of the reduced spontaneous curvature as a function of the reduced volume. To achieve this, a starting value of reduced spontaneous curvature c_0 and a starting set of values for the design parameters are chosen. This allows finding a stationary state for a given value of the reduced volume v. The optimization is repeated for a new value of v using the previously found stationary state as a

Fig. 7.8 Sample shapes
vesicles obtained during
optimization for $c_0 = 0.01$,
(**a**) $v = 0.581$, (**b**) $v = 0.789$,
(**c**) $v = 0.865$

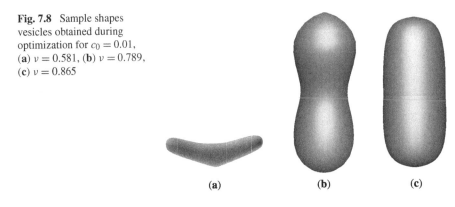

(**a**)   (**b**)   (**c**)

starting point for the new optimization. Thus, for a given value of c_0 starting from $v = 0.5$ the optimization repeated until $v = 1.0$ has been reached.

Figures 7.8 and 7.9 show the results of sample vesicle shapes of different volumes obtained for $c_0 = 0$ and $c_0 = 3.0$, respectively. These results can be validated against the various test cases reported in [4].

7.6 Conclusions

The purpose of this chapter was to present a methodology for design for function based on automatic design optimization using PDE formulation, enabling efficient shape definition and shape parametrization. In this chapter, we have seen several examples which demonstrate how the methodology can be used for automatic design optimization. These examples clearly demonstrate that the proposed approach to shape parametrization when combined with a standard method for numerical opti-

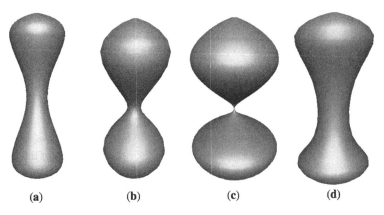

(**a**)   (**b**)   (**c**)   (**d**)

Fig. 7.9 Sample shapes vesicles obtained during optimization for $c_0 = 3.0$, (**a**) $v = 0.812$, (**b**) $v = 0.689$, (**c**) $v = 0.711$, (**d**) $v = 0.762$

mization is capable of setting up automatic design optimization problems, allowing practical design for function to be feasible.

References

1. Bates DM, Watts DG (1988) Nonlinear regression and its applications. Wiley, New York
2. Canham PB (1970) The minimum energy of bending as a possible explanation of the biconcave shape of the human red blood cell. J Theor Biol 26:61–81. doi:10.1016/S0022-5193(70)80032-7
3. Chapelle D, Bathe K (2010) The finite element analysis of shells—fundamentals. Springer, Berlin
4. Gennis RB (1989) Biomembranes: molecular structure and function. Springer, New York
5. Greig DM (1980) Optimisation. Longman, London
6. Hinton E, Owen DRJ (1984) Finite element software for plates and shells. Pineridge Press, Swansea
7. Kirkpatrick S, Gellat D Jr, Vecchi MP (1983) Optimisation by simulated annealing. Science 220(4598):671–680
8. Laarhoven PJM, Aarts EHL (1987) Simulated annealing: theory and applications mathematics and its applications. Reidel, Holland
9. Metropolis N, Rosenbluth AW, Rosenbluth MN, Teller AH, Teller E (1953) Equations of state calculations by fast computing machines. J Chem Phys 21(6):1087–1092. doi:10.1063/1.1699114
10. Struik D (1961) Lectures on classical differential geometry. Addison-Wesley, Reading

Chapter 8
Other Applications

Abstract This chapter presents a number of other application areas (which have not been discussed in previous chapters) which can benefit from using PDEs for geometric design. Particularly, in this chapter we show how PDEs can be effectively used for animation, data representation and compression. Furthermore, we discuss an emerging area of research where PDE based geometric design is being related to traditional spline based techniques.

8.1 Use of PDEs for Generating Time Dependent Geometry and Animation

In this section, we discuss how PDEs can be used to generate time dependent geometry. Particularly, we show how the boundary functions which define the PDE can be given as functions of time. This enables the geometry of a shape to be controlled and further manipulated according to a certain set of rules which have physically important meaning [1].

The essential idea behind creating efficient shape parametrization of time dependent geometry is to take these design parameters to be functions of time, thus enabling a user to create time-dependent geometric models whose motion can be controlled using a handful of design parameters.

An important point noteworthy here is that the use of analytic solution techniques for solving PDEs enables fast generation of the geometry, which usually takes a fraction of a second, thus enabling real-time animations. One can essentially take an initial geometry and introduce shape parameters on the boundary conditions and the spine, where the parameters are made time-dependent in order to simulate the animation process. Thus, given an initial parametrization, the geometry is re-generated for each time step of the animation process where the time-dependent parameters are calculated at each stage and the corresponding PDE surfaces are generated. Noticeably, this process has the advantage over existing animation techniques such as key framing in that the animation process is essentially controlled by the animator via the time dependent shape parameters.

H. Ugail, *Partial Differential Equations for Geometric Design*,
DOI 10.1007/978-0-85729-784-6_8, © Springer-Verlag London Limited 2011

Fig. 8.1 The generic shape
of a left ventricle of a human
heart. (**a**) The position
boundary curves. (**b**) The
resulting shape of the left
ventricle generated using a
combination of four separate
PDE surface patches

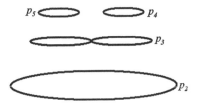

(a)

(b)

8.1.1 Modeling the Time Dependent Geometry of a Human Heart

In this section, we describe how one can model a ventricle of a human heart where
the beating of the heart is simulated. Figure 8.1 shows a representation of the left
ventricle of the heart. This particular geometry involves four separate PDE surface patches which are appropriately joined with common boundaries. Figure 8.1(a)
shows the position boundary curves defining the entire ventricle shape. For the sake
of bringing clarity to the illustration, the position boundary curves are only shown
here, thus omitting the corresponding derivative curves from the figure. The derivative boundary curves can be generated by means of transforming the corresponding position curves and in the cases when the surface patches meet each other, the
derivative curves are taken in a manner such that tangent plane continuity is ensured
between the patches. Thus, as shown in Fig. 8.1(b), the four surface patches which
make up the left ventricle can be created using separate surface patches between the
curves p_1 and p_2, p_2 and p_3, p_3 and p_4, and p_3 and p_5. Note that the curve p_1 can

be initially taken to be a circle where its radius can be then effectively reduced to zero in order to create a point in $\mathbf{R}^3$.

In order to simulate the time-dependent motion of the left ventricle, we utilize the information provided in [6] describing the changes in the ventricular shape in a heart beat. Essentially, a heart beat involves three components of motion, namely, axial contraction, radial contraction and twisting motion. To model these three modes of motion with realistic changes in volume and dimension of the heart, we define the time dependent parameters based on the boundary conditions shown in Fig. 8.1(a). Thus, taking t to be the time, the analytic forms of these boundary conditions are given as follows:

$$p_2(v, t) = \left(x_2 + r_2(t)\cos\left(v + \alpha(t)\right), y_2 + r_2(t)\sin\left(v + \alpha(t)\right), z_2 - z(t)\right), \quad (8.1)$$

$$p_3(v, t) = \left(x_3 + r_3(t)\cos\left(v + \alpha(t)\right)\sin\left(v - \alpha(t)\right),\right.$$
$$\left. y_3 + r_3(t)\sin\left(v + \alpha(t)\right), z_3 - z(t)\right), \quad (8.2)$$

$$p_4(v, t) = \left(x_4 + r_4\cos\left(v + \alpha(t)\right), y_4 + r_4\sin\left(v + \alpha(t)\right), z_{top} - z(t)\right), \quad (8.3)$$

$$p_5(v, t) = \left(x_5 + r_5\cos\left(v + \alpha(t)\right), y_5 + r_5\sin\left(v + \alpha(t)\right), z_{top} - z(t)\right), \quad (8.4)$$

where

$$z(t) = z_{ch}\sin^2(t\pi/t_m), \quad (8.5)$$

$$r_2(t) = r_2\left(1 - r_{ch}\sin^2(t\pi/t_m)\right), \quad (8.6)$$

$$r_3(t) = r_3\left(1 - r_{ch}\sin^2(t\pi/t_m)\right), \quad (8.7)$$

and

$$\alpha(t) = \alpha_{ch}\sin^2(t\pi/t_m). \quad (8.8)$$

The parameter t_m is the time period of a typical cardiac cycle which can be taken to be 0.8. The function $z(t)$ controls the axial contraction where the parameter r_{ch} can be taken as 0.7. Similarly, the function $r(t)$ controls the radial contraction where the parameter r_{ch} can be taken as 0.3. Finally, the twisting motion is generated via the parameter $\alpha(t)$ with α_{ch} accounting for the amount of twist in the ventricle. Here the value of α_{ch} was taken to be $\pi/4$.

With the above parameters describing the dynamics of the heart, it is seen that realistic animations accounting for the changes in the volume and dimension of the ventricle for a typical cardiac cycle can be generated in real-time. Figure 8.2 shows a series of images illustrating the beating of the generic shape of the left ventricle using the time-dependent shape parameters discussed here.

8.1.2 Facial Animation

Previously in Chap. 4, we have shown how facial geometry can be represented using PDEs. Here the input to the systems is a set of curves resembling facial geometry.

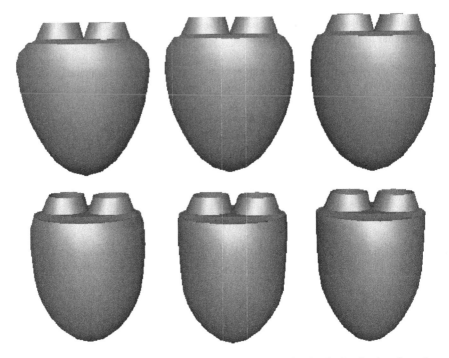

Fig. 8.2 A sequence of shapes showing a typical cardiac cycle simulated using the time-dependent shape parameters introduced on the boundary curves of the corresponding PDE surface patches

This set of curves can be either drawn or extracted form a three-dimensional mesh model, which in general are obtained from using a 3D laser scanner. Thus, given a 3D scan model of a human face, the standard procedure here is to extract a set of curves. Each of the curves can be appropriately parameterized so that they can be used as boundary conditions to solve a PDE.

An efficient parameterizations of a PDE model of a human face can be a very useful tool for facial animation since an intuitive and controlled manipulation of the generating curves can lead to realistic facial expressions [4]. Moreover, its parametric formulation would free valuable storage resources since the animation frames may be produced as needed. It is also a good alternative for producing animation sequences for applications with limited computing resources such as mobile phones or virtual communication since the PDE method can be used obtain simplified versions of very complicated models without losing a great deal of detail. Mathematical relations for some of the most common facial expressions can be in the form of simple mathematical functions such as translations and sinusoidal functions which can model expressions such as smiling, frowning and eyebrow raising.

Figure 8.3 presents various PDE geometries representing different facial expressions. These expressions include a left side smile, a right smile with a right eyebrow raise together with unilateral left and right frowns.

Another application that arises from this is the so-called motion re-targeting whereby an existing mesh model can be animated using PDE generated template

Fig. 8.3 PDE surface representations of different facial expressions

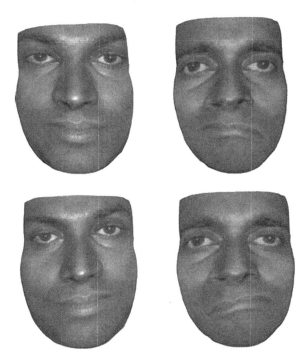

faces. That is, animation sequences can be transferred form a PDE-based model to any mesh model semi-automatically as seen in Fig. 8.4 in which facial expressions generated using PDE surfaces are then transferred to mesh models.

8.1.3   Cyclic Animation

Another area that has benefited from the use of the PDEs for geometric design is cyclic animation [2]. Here we provide two examples to illustrate this.

8.1.3.1   Human Body Animation

Rig-based animations are those in which a rigid skeleton is manipulated either manually or automatically to create different poses. An analogy between the rig and the human body is perfectly valid since both are responsible for providing support and define the position of an object. The layer covering the rig is known as skin (again, note the analogy with the human body) and this layer is the one that is visible when the object is rendered on the screen. The PDE method can be employed here to produce the skin of a rig and then can be used for simulating a number of cyclic activities by humans. These activities include: crawling, walking, running, cycling,

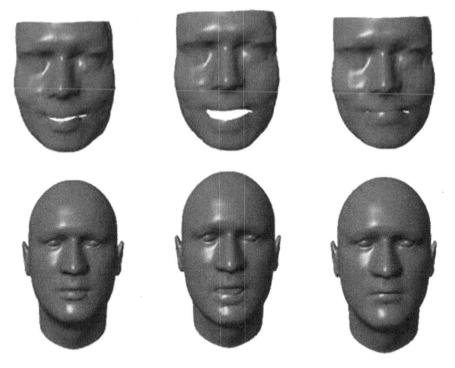

Fig. 8.4 An example of motion re-targeting from PDE based animation to a mesh model

swimming and dancing. All of these activities can be performed by simply adjusting a rig that could be built using standard modeling and animation software such
as Maya from Autodesk. The adjustments consist of changing the positions of the
structures associated with the arms and legs accordingly and changing the values
of a set of parameters which control the speed, frequency height and length of the
movement. Thus, the PDE method can be employed as a means of skinning such
a surface at every animation cycle. This is achieved by placing a set of boundary
curves outlining the silhouette of a human body accordingly and then computing
the resulting PDE surface representation for each of the animation frames. Note that
this technique is entirely independent of the skinning technique. This technique can
also be employed to animate fictitious biped characters which, in general, are represented by very detailed (and hence huge) geometric mesh models. Figure 8.5(a)
shows the generating boundary curves together with the same views of the resulting
PDE geometry. Figure 8.6(b) shows how the boundary curves are attached to the
rig; this has been created using Maya software.

8.1.3.2 Spine-Based Animation for Modeling Fish Locomotion

As discussed earlier, a mathematical property inherent to the PDE method is that of
identifying a spine with one of the terms of the Fourier series associated with the

Fig. 8.5 (**a**) Boundary curves
for generating the PDE
geometry representing the
human body. (**b**) Illustration
of how the generating curves
are attached to the rig

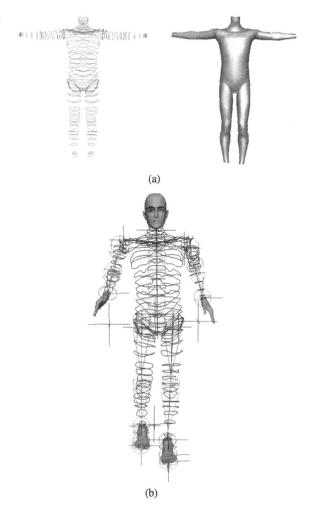

(a)

(b)

analytic solution to the PDE as described in Chap. 4. A framework for simulating
the oscillatory movement observed in most of the aquatic animals can be developing
by modeling their movement using PDEs where the spine is the main driver of the
animation [7]. The procedure to achieve this can be described as follows. First, the
spine and the radial component of the PDE surface representing the fish are identified. Second, the spine is manipulated by means of a sinusoidal function. Finally, the
radial component is added to obtain the corresponding surface representation. Note
that spine-based animations are ideal to be transferred to different geometric models
without spending a great deal of additional effort since the movement is applied to
a feature that both models have in common.

In the case of fish locomotion, the spine can be described as

$$Sp_i(u, t) = Sp_{original} + \Omega(u)\cos(\alpha u + \phi)\sin(2\pi \omega t_i), \tag{8.9}$$

Fig. 8.6 PDE surface
representations of two
different poses over a dancing
cycle obtained after skinning
a rig

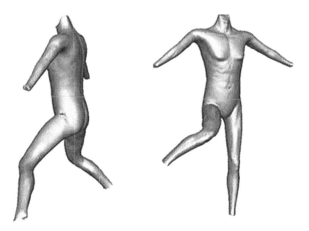

where $\Omega(u)$ determines the amplitude and depends upon u, ϕ represents the phase, α denotes the wave number and ω regulates the frequency of the undulatory movement. The subscript i determines the frame for which the animation cycle, and $t_i \in [0, 1]$ is the time associated with each respective frame. Notice that u also requires to be normalized so that its value varies from 0 to 2π.

It is also worth mentioning that for the time being, only the animation of the segment of the spine representing the length of the fish is considered (8.9), leaving the fins and the tail of the fish without movement. The parameters associated with this equation can be adjusted so that different types of fish, varying from water snakes to dolphins, can be animated using the same formulation. Figure 8.7 represents the position of the spine corresponding to different times in the animation cycle. Some frames representing the swimming cycle of a water snake are shown in Fig. 8.8, the left side of the figure shows the resulting PDE surface representing the water snake whilst the right presents the animated spine corresponding to each frame. Different types of fish have been animated using point-to-point correspondence, an example of this can be seen in Fig. 8.9.

8.2 Use of PDEs for Data Representation and Compression

In real-life, complex geometric shapes are often represented in terms of large polygon meshes or point sets. This requires a considerable amount of data storage and memory for handling such data. One technique to reduce the amount of data is to represent the data by means of approximation in terms of mathematical functions. Common methods for this include the use of splines such as NURBS. However, such methods for data representation require storing a relatively large set of control points which are then used by the interpolation procedures during the run time to produce the polygons.

On the other hand, for model representation, the boundary-value approach adopted by the PDE seems to be far more efficient. The basic idea here consists

Fig. 8.7 Evolution of the spine over a cycle. The *horizontal line* represents both the initial and final position of the spine, whereas the remaining outline its position at different values of *t*

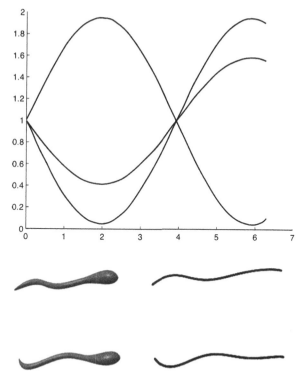

Fig. 8.8 Different animation frames of a swimming cycle of a PDE surface representation of a water snake (*left*) and its corresponding spine (*right*)

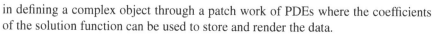

in defining a complex object through a patch work of PDEs where the coefficients of the solution function can be used to store and render the data.

Thus, in order to represent a complex polygonal object with irregular and sharp geometric details in an efficient way, one can make use of a patchwise PDE method where the configuration of the PDE patches are matched to the complexity of the geometric model being approximated [5].

An important point to bear in mind here is that, due to the inherent nature of the analytic solution utilized here to represent the patches, once represented in PDE format, within each patch the rendering can be performed at arbitrary level of surface resolution.

Figure 8.10 shows the curve patches that can be used to represent a sphere. The usual procedure employed to extract curves is to utilize a template which consists of a low resolution approximation closely matching the object. Figure 8.10(a) contains the template boundary patches extracted from the low resolution sphere model and the final PDE patches. The images in Fig. 8.10(b) shows the final PDE surface at

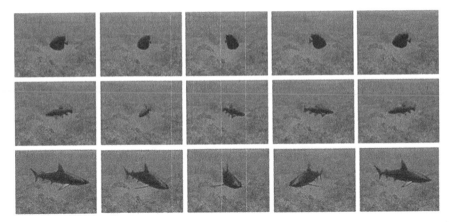

Fig. 8.9 Examples of realistic fish locomotion using PDE based motion re-targeting

Fig. 8.10 Curve set for representing the sphere. (**a**) The complete curve set for the template boundary curves that represents sphere model. (**b**) Original sphere (on the *left*) and two different subdivision levels generated using PDE approximation of the sphere

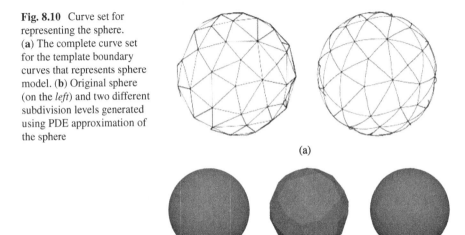

two different subdivision levels compared with the original sphere model (on the left).

8.3 Biharmonic Bézier Surfaces

In Chap. 3, we briefly hightailed that both the Harmonic and Biharmonic operators are widely used in many application areas. For example, the Harmonic operator, otherwise known as the Laplacian, is associated with a wide range of physical problems such as gravity, electromagnetism as well as fluid flows. Similarly, the Biharmonic is also associated with a variety of physical problems such as tension in elastic membranes and the study of stress and strain in physical structures.

From a geometric design point of view, these operators have found their way into geometric design itself, e.g. the PDEs applied to geometric design we have discussed.

Recent research work has also tried to find the connection between the elliptic PDE operators and traditional splines, for example, in generating Bézier surfaces which verify both the Harmonic and Biharmonic Bézier conditions [3].

Harmonic surfaces are related to minimal surfaces: surfaces that minimize the area among all surfaces with prescribed border. The relation is the following: If a patch **x** is isothermal then the surface it represents is minimal if and only if it is harmonic. The main result from the theory of minimal surfaces states that, under certain conditions, given the boundary there is a unique minimal surface prescribed by that boundary.

Furthermore, similar to minimal surfaces, the knowledge of the boundary of a Biharmonic Bézier surface fully determines the entire surface. This is to be compared with the corresponding result for Harmonic Bézier surfaces where the knowledge of two opposite boundary curves of the Harmonic Bézier surface fully determines the surface.

Given a quadratic net of points in $\mathbf{R}^3$, $\{P_{ij}\}_{i,j=0}^n$, the associated Bézier surface, $\mathbf{x} : [0, 1] \times [0, 1] \to \mathbf{R}^3$, is harmonic, provided for any $i, j \in \{1, \ldots, n\}$,

$$
\begin{aligned}
0 = {} & P_{i+2,j} a_{n,i,0} + P_{i+1,j}(a_{n,i-1,1} - 2a_{n,i,0}) + P_{i-1,j}(a_{n,i-1,1} - 2a_{n,i-2,2}) \\
& + P_{i-2,j} a_{n,i-2,2} + P_{i,j+2} a_{n,j,0} + P_{i,j+1}(a_{n,j-1,1} - 2a_{n,j,0}) \\
& + P_{i,j-1}(b a_{n,j-1,1} - 2a_{n,j-2,2}) + P_{i,j-2} a_{n,j-2,2} \\
& + P_{ij}(a_{n,i,0} - 2a_{n,i-1,1} + a_{n,i-2,2} + a_{n,j,0} - 2a_{n,j-1,1} + a_{n,j-2,2}). \quad (8.10)
\end{aligned}
$$

This states that the Harmonic condition implies that some of the control points can be expressed as linear combinations of the other control points, i.e. the first and last rows of control points fully determine the Harmonic Bézier surface.

Similarly, given a control net in $\mathbf{R}^3$, $\{P_{ij}\}_{i,j=0}^{n,m}$, the associated Bézier surface, $\mathbf{x} : [0, 1] \times [0, 1] \to \mathbf{R}^3$, is biharmonic, provided for any $i \in \{1, \ldots, n\}$ and $j \in \{1, \ldots, m\}$

$$
\sum_{k=0}^{4} b_{n,i-k,k} \Delta^{4,0} P_{i-k,j} + 2 \sum_{k,\ell=0}^{2} a_{n,i-k,k} a_{m,j-\ell,\ell} \Delta^{2,2} P_{i-k,j-\ell}
$$

$$
+ \sum_{\ell=0}^{4} b_{m,j-\ell,\ell} \Delta^{0,4} P_{i,j-\ell}, \quad (8.11)
$$

where, for $i \in \{0, \ldots, n - 2\}$

$$
a_{ni0} = (n - i)(n - i - 1),
$$

$$
a_{ni1} = 2(i + 1)(n - i - 1),
$$

$$
a_{ni2} = (i + 1)(i + 2),
$$

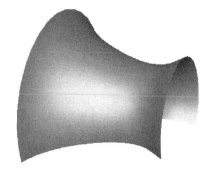

Fig. 8.11 A typical
Biharmonic Bézier surface
where the edges of the
surface patch define Bézier
curves which are taken as the
boundary conditions

and $a_{nik} = 0$ otherwise, and for $i \in \{0, \ldots, n-4\}$

$$b_{ni0} = (n-i)(n-i-1)(n-i-2)(n-i-3),$$

$$b_{ni1} = 4(i+1)(n-i-1)(n-i-2)(n-i-3),$$

$$b_{ni2} = 6(i+1)(i+2)(n-i-2)(n-i-3),$$

$$b_{ni3} = 4(i+1)(i+2)(i+3)(n-i-3),$$

$$b_{ni4} = (i+1)(i+2)(i+3)(i+4),$$

and $b_{nik} = 0$ otherwise.

Figure 8.11 shows a typical Biharmonic Bézier surface where the edges of the surface patch define Bézier curves which are taken as the boundary conditions.

8.4 Conclusions

In this chapter, a number of applications which make use of PDEs for geometric design are discussed. Examples of areas discussed include animation, data modeling and compression. This chapter should have provided a flavor of the various application domains as well as potential application domains related to geometric design whereby PDEs can play a crucial role.

References

1. Castro CG, Ugail H, Willis P, Palmer I (2008) A survey of partial differential equations in geometric design. Vis Comput 24(3):213–225. doi:10.1007/s00371-007-0190-z
2. Castro G, Athanasopoulos M, Ugail H (2010) Cyclic animation using partial differential equations. Vis Comput 26(5):325–338. doi:10.1007/s00371-010-0422-5
3. Monterde J, Ugail H (2006) A general 4th-order PDE method to generate Bézier surfaces from the boundary. Comput Aided Geom Des 23(2):208–225. doi:10.1016/j.cagd.2005.09.001
4. Sheng Y, Willis P, Castro G, Ugail H (2009) PDE-based facial animation: making the complex simple, In: Advances in visual computing, part II. Lecture notes in computer science (LNCS), vol 5359. Springer, Berlin, pp 723–732

5. Sheng Y, Sourin S, Gonzalez Castro G, Ugail H (2010) A PDE method for patchwise approximation of large polygon meshes. Vis Comput 26(6–8):975–984. doi:10.1007/s00371-010-0456-8

6. Smith JJ, Kampine JP (1990) Circulatory physiology, the essentials. Williams and Wilkins, Baltimore

7. Ugail H (2003) On the spine of a PDE surface. In: Wilson MJ, Martin RR (eds) Mathematics of surfaces X. Springer, Berlin, pp 366–376

8. Ugail H (2004) Spine based shape parameterisation for PDE surfaces. Computing 72:195–206. doi:10.1007/s00607-003-0057-8

9. Ugail H, Bloor MIG, Wilson MJ (1999) Techniques for interactive design using the PDE method. ACM Trans Graph 18(2):195–212. doi:10.1145/318009.318078

Chapter 9
Conclusions

This book introduced to the reader the use of partial differential equations for geometric design which has been an important and fast moving field and has many computer based application areas ranging from computer based engineering to computer animation.

Common geometric design tools available today face many underlying problems. These include the lack of efficient computer-based techniques to create a satisfactory design from 'scratch', the difficulty a designer faces when interactively manipulating an existing geometry model, and the problem of generating an optimal design to serve a specific purpose. The root of this problem is the lack of appropriate mathematical tools which can represent the geometry as well as facilitate the calculation of the necessary functional aspects of the object one intends to design.

One of the requirements for such a geometric design tool is that it should possess as much flexibility as possible. In other words, it is crucial to have methods and techniques which are capable of producing a wide range of alternative shapes using a minimum number of design parameters. Furthermore, for the purpose of maintaining consistency, it would be desirable to use the same design parameters for both analysis and optimization.

In this book, it has been shown that the use of elliptic PDEs is one of possible mechanisms through which complex geometry can be generated, manipulated, parameterized and furthermore optimized for specific needs. The positive features of the geometry generated using PDEs is the ability to generate fair surfaces, to maintain continuity between adjacent surface patches and the global control that the design parameters have upon the shape of the surface. Thus, PDEs are capable of parameterizing complex shapes in terms of a small set of design parameters whilst maintaining sufficient flexibility in the range of shapes generated.

Hence, in this book we describe the geometric design using PDEs whereby the design methodology can be divided into three categories: the interactive definition and parametrization of the geometry, the integration of the design with analysis using the original parametrization and the design optimization for a chosen design merit function with some imposed constraints. Thus, the same parametric model generated using a given PDE can be used to create and manipulate geometry, to

H. Ugail, *Partial Differential Equations for Geometric Design*,
DOI 10.1007/978-0-85729-784-6_9, © Springer-Verlag London Limited 2011

enable the appropriate boundary conditions for analysis to be set up and to enable the imposition of design constraints during optimization.

Appendix
Maple Code to Generate a Surface Patch for the Biharmonic Equation

The Maple code below can be used to generate a simple PDE surface patch for the Biharmonic Equation. Here the Biharmonic Equation is solved analytically based on the solution discussed in Chap. 4. Here we assume the boundary conditions are given in terms of simple analytic functions which can be represented in terms of a finite Fourier series. In this particular case, it is assumed that the boundary conditions are described using the first mode of the Fourier series. Note that the resulting surface is parametric in which suitable boundary conditions are supplied for each Cartesian coordinate.

The code can be easily adapted for more complex cases whereby boundary conditions using more than one mode of the Fourier series can be prescribed or even boundary curves defined by discrete set of points can also be utilized, in which case the method described in Chap. 4 to obtain a Finite Fourier series from a discrete set of points can be used.

```
> restart;
> with(plots):
> a:=1.1;
> A[0]:=a[0]+a[1]*u + a[2]*u^2 + a[3]*u^3;
>
A[n]:=a[11]*exp(a*u)+a[12]*u*exp(a*u)+a[13]*exp(-a*u)+a[14]*u*ex
p(-a*u);
>
B[n]:=b[11]*exp(a*u)+b[12]*u*exp(a*u)+b[13]*exp(-a*u)+b[14]*u*ex
p(-a*u);
> pde:=A[0] + (A[n]*cos(v)+B[n]*sin(v));
>
>
> dA[0]:= diff(A[0],u):
> dA[n]:= diff(A[n],u):
> dB[n]:= diff(B[n],u):
> eqA01 := subs(u=0,A[0])=u_0:
> eqA02 := subs(u=1,A[0])=u_1:
> eqA03 := subs(u=0.2,A[0])=du_0:
> eqA04 := subs(u=0.8,A[0])=du_1:
> sol1:=solve({eqA01, eqA02, eqA03, eqA04}, {a[0],a[1], a[2],
a[3]});
>
```

H. Ugail, *Partial Differential Equations for Geometric Design*,
DOI 10.1007/978-0-85729-784-6, © Springer-Verlag London Limited 2011

```
>
> eqAn1 := subs(u=0,A[n])=Au_0: simplify(eqAn1):
> eqAn2 := subs(u=1,A[n])=Au_1: simplify(eqAn2):
> eqAn3 := subs(u=0.2,A[n])=Adu_0: simplify(eqAn3):
> eqAn4 := subs(u=0.8,A[n])=Adu_1: simplify(eqAn4):
>
>
>
> sol2:=solve({eqAn1, eqAn2, eqAn3, eqAn4}, {a[11],a[12], a[13],
a[14]}):
>
> eqBn1 := subs(u=0,B[n])=Bu_0: simplify(eqBn1):
> eqBn2 := subs(u=1,B[n])=Bu_1: simplify(eqBn2):
> eqBn3 := subs(u=0.2,B[n])=Bdu_0: simplify(eqBn3):
> eqBn4 := subs(u=0.8,B[n])=Bdu_1: simplify(eqBn4):
>
> sol3:=solve({eqBn1, eqBn2, eqBn3, eqBn4}, {b[11],b[12], b[13],
b[14]}):
>
>
>




> #z component
> u_0:=0.0: u_1:=1: du_0:=0.5: du_1:=0.8:
> Au_0:=0: Au_1:=0: Adu_0:=0: Adu_1:=0:
> Bu_0:=0: Bu_1:=0: Bdu_0:=0: Bdu_1:=0:
> z:=subs(sol1,sol2, sol3, pde);
>
> spine := subs(sol1, A[0]);
>
> #x component
> u_0:=0: u_1:=0: du_0:=0: du_1:=0:
> Au_0:=0.5: Au_1:=2.0: Adu_0:=0.5: Adu_1:=2.0:
> Bu_0:=0.0: Bu_1:=0.0: Bdu_0:=0: Bdu_1:=0:
> x:=subs(sol1,sol2, sol3,pde):
>
>
> #y component
> u_0:=0: u_1:=0: du_0:=0: du_1:=0:
> Au_0:=0: Au_1:=0: Adu_0:=0: Adu_1:=0:
> Bu_0:=0.5: Bu_1:=2.0: Bdu_0:=0.5: Bdu_1:=2.0:
> y:=subs(sol1,sol2, sol3,pde):
>
>
> surface := [x,y,z]:
> spi := [0,0,spine]:
> plot3d({surface, spi},u=0..1,v=0..2*Pi,axes=NORMAL,
shading=zhue);
>
```

Index

H. Ugail, *Partial Differential Equations for Geometric Design*,
DOI 10.1007/978-0-85729-784-6, © Springer-Verlag London Limited 2011